AA004074

GLOBAL HEALTH 2018

The Seventh International Conference on Global Health Challenges

November 18 - 22, 2018

Athens, Greece

GLOBAL HEALTH 2018 Editors

Hassan Khachfe, Lebanese International University, Lebanon

Birgit Gersbeck-Schierholz, Leibniz Universität Hannover, Germany

Printed from e-media with permission by:

Curran Associates, Inc.
57 Morehouse Lane
Red Hook, NY 12571

Some format issues inherent in the e-media version may also appear in this print version.

Printed by Curran Associates, Inc. (2019)

International Academy, Research, and Industry Association (IARIA)
412 Derby Way
Wilmington, DE 19810

Phone: (408) 893-6407
Fax: (408) 527-6351

petre@iaria.org

Additional copies of this publication are available from:

Curran Associates, Inc.
57 Morehouse Lane
Red Hook, NY 12571 USA
Phone: 845-758-0400
Fax: 845-758-2633
Email: curran@proceedings.com
Web: www.proceedings.com

Table of Contents

GLOBAL HEALTH 2018

Forward

The Seventh International Conference on Global Health Challenges (GLOBAL HEALTH 2018), held between November 18, 2018 and November 22, 2018 in Athens, Greece, took a global perspective on population health, from national to cross-country approaches, multiplatform technologies, from drug design to medicine accessibility, everything under mobile, ubiquitous, and personalized characteristics of new age population.

Recent advances in technology and computational science influenced a large spectrum of branches in approaching population health. Despite significant progresses, many challenges exist, including health informatics, cross-country platforms interoperability, system and laws harmonization, protection of health data, practical solutions, accessibility to health services, and many others. Along with technological progress, personalized medicine, ambient assistance and pervasive health complement patient needs. A combination of classical and information-driven approach is developing now, where diagnosis systems, data protection mechanisms, remote assistance and hospital-processes are converging.

The conference had the following tracks:

- Health and Wellness Informatics
- Challenges
- Alternative
- Global

We take here the opportunity to warmly thank all the members of the GLOBAL HEALTH 2018 technical program committee, as well as all the reviewers. The creation of such a high quality conference program would not have been possible without their involvement. We also kindly thank all the authors that dedicated much of their time and effort to contribute to GLOBAL HEALTH 2018. We truly believe that, thanks to all these efforts, the final conference program consisted of top quality contributions.

We also gratefully thank the members of the GLOBAL HEALTH 2018 organizing committee for their help in handling the logistics and for their work that made this professional meeting a success.

We hope that GLOBAL HEALTH 2018 was a successful international forum for the exchange of ideas and results between academia and industry and to promote further progress in the domain of global health. We also hope that Athens, Greece, provided a pleasant environment during the conference and everyone saved some time to enjoy the historic charm of the city.

GLOBAL HEALTH 2018 Chairs

GLOBAL HEALTH Steering Committee

Hassan Khachfe, Lebanese International University, Lebanon
Anthony Glascock, Drexel University - Philadelphia, USA

Shunji Shimizu, Kyushu University Hospital, Japan
Matthieu-P. Schapranow, Hasso Plattner Institute, Germany
Chun-Hsi Huang, University of Connecticut, USA
Vladislav Rajkovič, University of Maribor, Slovenia
Carsten Röcker, Fraunhofer Application Center Industrial Automation (IOSB-INA), Germany
Lisette Van Gemert-Pijnen, University of Twente - Enschede, The Netherlands
Marike Hettinga, Windesheim University of Applied Sciences, The Netherlands

GLOBAL HEALTH Technology/Research Advisory Committee

Liang Zhan, Department of Electrical and Computer Engineering, University of Pittsburgh, USA
Flora Malamateniou, University of Piraeus, Greece
Da-Wei Wang, Institute of Information Science | Academia Sinca, Taiwan
Sten Hanke, IPN Instituto Pedro Nunes, Portugal | AIT Austrian Institute of Technology GmbH, Austria
Liangjiang Wang, Clemson University - Greenwood, USA

GLOBAL HEALTH 2018
Committee

GLOBAL HEALTH Steering Committee

Hassan Khachfe, Lebanese International University, Lebanon
Anthony Glascock, Drexel University - Philadelphia, USA
Shunji Shimizu, Kyushu University Hospital, Japan
Matthieu-P. Schapranow, Hasso Plattner Institute, Germany
Chun-Hsi Huang, University of Connecticut, USA
Vladislav Rajkovič, University of Maribor, Slovenia
Carsten Röcker, Fraunhofer Application Center Industrial Automation (IOSB-INA), Germany
Lisette Van Gemert-Pijnen, University of Twente - Enschede, The Netherlands
Marike Hettinga, Windesheim University of Applied Sciences, The Netherlands

GLOBAL HEALTH Technology/Research Advisory Committee

Liang Zhan, Department of Electrical and Computer Engineering, University of Pittsburgh, USA
Flora Malamateniou, University of Piraeus, Greece
Da-Wei Wang, Institute of Information Science | Academia Sinca, Taiwan
Sten Hanke, IPN Instituto Pedro Nunes, Portugal | AIT Austrian Institute of Technology GmbH, Austria
Liangjiang Wang, Clemson University - Greenwood, USA

GLOBAL HEALTH 2018 Technical Program Committee

Harshavardhan Achrekar, University of Massachusetts - Lowell, USA
Faisal A. Alkhaldi, Tennessee Technological University, USA / Prince Mohammed Medical City, Saudi Arabia
Kyota Aoki, Utsunomiya University, Japan
Marzieh Ayati, Case Western Reserve University, Cleveland, USA
Thierry Badard, Centre for Research in Geomatics | Laval University, Quebec, Canada
Panagiotis D. Bamidis, Aristotle University of Thessaloniki, Greece
Arvind Bansal, Kent State University, USA
Pedro Barahona, Universidade Nova de Lisboa, Portugal
Debmalya Barh, Institute of Integrative Omics and Applied Biotechnology - IIOAB, India
Veselka Boeva, Blekinge Institute of Technology, Sweden
Ann Borda, The University of Melbourne, Australia
Alessio Bottrighi, DiSIT - Computer Science Institute - University of Eastern Piedmont, Italy
Pierrette Bouillon, Geneva University, Switzerland
Rui Camacho, University of Porto, Portugal
Ozgu Can, Ege University, Turkey
Antonio Celesti, University of Messina, Italy

Dechang Chen, Uniformed Services University of the Health Sciences, USA
Chih-Lin Chi, University of Minnesota, USA
Meyrick Chow, School of Nursing | Tung Wah College, Hong Kong
James J. Cimino, Informatics Institute | Medicine University of Alabama at Birmingham - School of Medicine, USA
Gouenou Coatrieux, Institut Mines Telecom - Telecom Bretagne, France
Steven A. Demurjian, University of Connecticut, USA
Behnam Dezfouli, Santa Clara University, USA
Christo El Morr, York University, Canada
Ivan Evgeniev Ivanov, Technical University of Sofia, Bulgaria
Dimeji Farri, Philips Research- North America, USA
Sebastian Fudickar, CvO University of Oldenburg, Germany
Angel M. Garcia-Pedrero, Universidad Politécnica de Madrid, Spain
Gersende Georg, French National Authority for Health, France
Anthony Glascock, Drexel University - Philadelphia, USA
Natalia Grabar, Université Lille 3, France
Adela Grando, Arizona State University, USA
David Greenhalgh, University of Strathclyde, UK
Lejla Gurbeta, International Burch University, Bosnia and Herzegovina
Sten Hanke, IPN Instituto Pedro Nunes, Portugal | AIT Austrian Institute of Technology GmbH, Austria
Tyler D. Harvey, Watson Foundation / Yale School of Public Health, USA
Marike Hettinga, Windesheim University of Applied Sciences, The Netherlands
Jeonghan Hong, Graduate School of Interdisciplinary Management - Ulsan National Institute of Science and Technology (UNIST), Korea
Pengwei Hu, The Hong Kong Polytechnic University, Hong Kong
Chun-Hsi Huang, University of Connecticut, USA
Ruyi Huang, E-Da Hospital & I-Shou University, Taiwan
Michal Huptych, Czech Institute of Informatics, Robotics, and Cybernetics (CIIRC) | Czech Technical University (CTU) in Prague, Czech Republic
Kamran Iqbal, University of Arkansas at Little Rock, USA
Ravi Janardan, University of Minnesota - Twin Cities, USA
Hassan Khachfe, Lebanese International University, Lebanon
Antonio Lanata', University of Pisa, Italy
Eva K. Lee, Georgia Institute of Technology, USA
Li-Hui Lee, National Taipei University of Nursing and Health Sciences, Taiwan
Gunther Lenz, Varian Medical Systems, USA
Lenka Lhotská, Czech Institute of Informatics, Robotics and Cybernetics | Czech Technical University in Prague, Czech Republic
Li Liao, University of Delaware, USA
Yuan Ling, Philips Research North America, USA
Iryna Lishchuk, Leibniz Universitaet Hannover | Institut für Rechtsinformatik, Germany
Igor Ljubi, University College for Applied Computer Engineering - Zagreb, Croatia
Jin Lu, University of Connecticut, USA

Pranjul Yadav, Criteo, USA
Erliang Zeng, University of South Dakota, USA
Liang Zhan, Department of Electrical and Computer Engineering, University of Pittsburgh, USA
Chi Zhang, University of Nebraska - Lincoln, USA
Peng Zhao, University of Missouri, USA
Qiang Zhu, The University of Michigan, USA

Copyright Information

Open-source Data Analysis and Machine Learning for Asthma Hospitalisation Rates

Laura Rooney, Chaloner Chute
Digital Health and Care Institute,
University of Strathclyde,
Glasgow, UK.
laura.rooney@dhi-scotland.com
chalonerchute@dhi-scotland.com

William J Buchanan, Adrian Smales
The Cyber Academy,
Edinburgh Napier University,
Edinburgh, UK.
w.buchanan@napier.ac.uk
a.smales@napier.ac.uk

Leigh-Anne Hepburn
Digital Health and Care Institute,
Glasgow School of Art,
Glasgow, UK.
leigh-anne.hepburn@dhi-scotland.com

***Abstract*—Long-term conditions in Scotland account for 80% of all GP consultations; they also account for 60% of all deaths in Scotland. Asthma and Chronic Obstructive Pulmonary Disease (COPD) are common long-term respiratory diseases [1]. Asthma is a heterogeneous disease, usually characterized by chronic airway inflammation. It is defined by the history of respiratory symptoms such as wheeze, shortness of breath, chest tightness and cough that vary over time and in intensity, together with variable expiratory airflow limitation [2]. So far, we know that there are many different things – such as viruses, allergens, and pollution – that cause asthma or trigger attacks but not why or how they do it. This paper outlines how an open source dataset can be used to estimate asthma hospitalisation rates and uses machine learning to predict these rates, within ±7.5%, and for an 86.67% success rate.**

I. Introduction

The prevalence of asthma continues to increase worldwide (it affects 1-18% of the global population), although non-communicable diseases such as Asthma are still not seen as a health care priority in many countries despite their impact. Asthma is a chronic inflammatory disease that presents as a long-term condition with severity varying from person to person. Globally, 300 million people are affected by asthma on a daily basis [3]. In 2015, the UK had the highest number of Asthma sufferers in Europe with 4.67 million people [4].

In the UK now, 1 in 11 people have asthma, including 1.1 million children and 4.3 million adults. Shockingly, every 10 seconds, someone in the UK has a potentially life-threatening asthma attack and each day three people in the world die from asthma [5]. The landmark International Study of Asthma and Allergies in Childhood (ISAAC) and the European Community Respiratory Health Survey (ECRHS) studies demonstrated substantial national variations in asthma prevalence, with evidence suggesting that the UK ranks as being one of the countries with the highest prevalence in the world [6]. Bringing these statistics down to a more regional level, Scotland has one of the highest prevalence of asthma in Europe with 9.51% of the population being affected by the condition (554,306 people) [3]. It is also predicted that by 2025, there will be almost 400 million asthma patients worldwide.

Scotland provides a good test case in analysing health and social care factors, especially where its local authorities and health boards often face different challenges in regard to health and well-being. While national statistics, such as SIMD (Scottish Index of Multiple Deprivation) exist [7], they do not give enough details on the factors which could influence a more in-depth analysis of contributory factors. This paper thus uses the publicly available data set from ScotPHO Profiles [8], and uses this to train towards the best machine learning models in order to predict the outcomes.

The methodology used in this paper to analyse asthma hospitalisation rates are to take an open source data set with a range of metrics and then:

- The metrics are correlated, using linear regression (with the Pearson correlation coefficient) against each other to see significant linkages.
- A machine is trained using 30% of the data and then to predicts from hospitalisation for asthma. In this way, the machine is being trained against three metrics within the data in order to determine the most significant factors in the matching.
- Each of the models are then assessed for the complete dataset, along with a success threshold. In this way, a success band within +/-7.5% between the minimum and maximum value is used. The top contenders for a successful match are then outlined and checked for their usefulness in showing contributory factors.
- Once all the models have been determined, the top machine learning models for the top 100 successful models are then selected and the variables from each on the most successful training models are then scored for their success. Those metrics which appear most often within the most successful models can then be defined as the most useful in predicting asthma hospitalisation rates.

II. Factors Influencing asthma prevalence

A. *Environmental factors*

Despite the considerable genetic contribution, for example, the presence of asthma and atopy in the family, most importantly the maternal history of asthma, there are likely to be several additional social and environmental factors involved in the exacerbation of asthma [9]. With most of the world's population living in urban areas, the environmental conditions whereby pollution continues to increase, have and will continue to have an influence on the rise in asthma prevalence. Air pollution – whether it's traffic fumes, smoke or dust particles – is an asthma trigger which is difficult to avoid in city environments. According to a survey carried out by Asthma UK, two-thirds of people reported that poor air quality makes their asthma worse.

B. Passive smoke

A study carried out to investigate the relationship between air quality and school absences found that exposure to cigarette smoke had a significant impact on current wheeze, use of services, and interference with physical activity. Among adolescents, smoke exposure was found to be a more important factor than deprivation in relation to symptoms, use of services, and impact on activities [10]. Passive smoking is a major global problem, causing 12,000 deaths each year in the UK and causing 2% of the current annual total deaths. These figures alone are enough to provide sufficient evidence to promote the prevention of passive smoking in the public domain [11].

C. Childhood obesity

It is also well documented that asthma presents its peak incidence in childhood. In the US, the age group with the highest percentage of persons with asthma from 2001 to 2015 was consistently age 5-14 years [4]. Obesity rates in children have increased dramatically across most English-speaking countries over the past decades and the impact on obesity on health has become a major global burden.

D. Poverty and asthma

A retrospective study carried out in West Midlands (England) to investigate the relationship between asthma admission rates, routes of admission and socioeconomic deprivation found that asthma admissions are strongly associated with deprivation in the community [12]. Asthma admission rates were higher in all age groups (except over 65s) for those from poorer districts. The age group with the most significant relationship between asthma admission rates and Townsend deprivation index was 0-4 years. Another study stipulates that higher hospital admission rates of asthmatics who are poor or belong to ethnic minorities may be due to the fact that these groups rely mainly on crisis management of the condition, [13] are under medicated [14], are under-users of primary care facilities, lack a planned crisis management, live in adverse environmental conditions in terms of asthma triggers such as smoking and cockroach exposure, or are exposed more frequently than other groups to psychosocial problems within the family and their community [15].

E. Asthma in the elderly

Although asthma has a high burden in children, the relative importance of asthma impact increases with old age and is also particularly apparent in the elderly, especially in women. Asthma reportedly affects 10% of the over 65 population who in general have lower lung function and greater symptom severity than young asthmatic patients. "late onset" asthma is first diagnosed after the age of 65 and is often a more severe phenotype, with less symptom-free days, and a higher requirement for oral corticosteroids [16].

III. METHODOLOGY

A. Open source data analysis

Population studies have been used within smoking and alcohol studies for many decades, such as in 1939 when Hermann Müller at Cologne Hospital in 1939 published his work on the linkage between smoking and cancer [17], and which was confirmed by Eberhard Schairer and Eric Schöniger [18]. Their work was then confirmed in the 1950s by a number of epidemiological studies, including Ernst Wynder and Evarts Graham [19]. Doll and Hill confirmed the effect by observing that smokers of 35 or more cigarettes per day had a 40.6 times increase in the odds of dying from lung cancer [20]. Since then smoking has been increasingly pinpointed within a number of ailments, including for COPD [21].

Smoking prevalence has also been studied and matched to differing demographics. This includes with [22] where researchers analysed 4,411 respondents aged 15 to 54 years and found that the smoking rates for no mental illness was 22.5%, while those with a lifetime mental illness was 34.8%. Along with this smoking is seen to be an increasing problem with poverty [23], especially as those who are wealthy are more likely to quit smoking than those in poverty [24].

B. Machine learning

In terms of asthma, machines learning is now playing an increasing role in creating expert systems for diagnosis. [25] used questionnaire data and clinical data to train a machine learning methods and used Context sensitive auto-associative memory neural network model (AMNN), a Backpropagation (BP) model, the C4.5 algorithm, a Bayesian Network (BN), and Particle Swarm Optimization (PSO). They found the accuracy of the methods varied between 81.17% and 84.16%, and that the accuracy of the AMNN and PSO methods were the most accurate, along with having excellent learning and diagnostic abilities.

[26] reviewed over five years of data and applied latent class analysis to distinguish asthma and wheezing subtypes in childhood. The data sets analysed included ALSPAC [27]; AMICS [28]; AMICS-Menorca [29]; CAPS [30]; CCCEH [31]; DARC [32]; ECRHSII [33]; EGEA2 [34]; GINIplus [35]; IoW; ISAAC phase II; Cohort; LISA; MAAS; MACS; MAS; MCS; PARIS; PASTURE; PIAMA; PIPO; SLAM; and WHEALS. Of the 34 data sets analysed, 16 included wheezing and coughing, 10 included atopic status, eight included growth patterns, and two included eczema. In the data sets, 22 related to children and only two related to adults, with the rest defined as unspecified. The cohort size was typically between 20 and 66.

In [36], the authors used a machine learning approach to analyse the pattern of IgE response (over time or to specific allergens) in order to identify atopic vulnerabilities related to the presence of asthma. They used data from the skin and IgE tests from childhood onwards and clustered the population into multiple atopic classes using unsupervised learning. These were trained against asthma-related data, such as for symptoms, hospitalizations, lung function and airway reactivity. Their results indicate four main classes: Multiple Early (10.6%); Multiple Late (16.2%); Dust Mite (4.5%); and Non-dust Mite (9.5%), along with a fifth class of No Latent Vulnerability (59.2%). The association for asthma was most strong for Multiple Early class, and which also showed a considerable link to lung function and airway reactivity. Along with this the work identified a highly significant increase in the risk of hospital admissions for wheeze/asthma after three years old, but only among children in the Multiple Early class.

IV. RESULTS

The results use a publicly available data set from ScotPHO Profiles [8], and where 56 metrics are used to train against (as outlined in Table III) and using the local authority region as the index value. If we select three variables to train against, one variable we will have 26,235 triplets to test, while four variables will give us 341,055 machine learning assessments (Table II). A benchmark of the time to check a model and to match against predicted values gives an estimated time of 0.4 seconds. Table II thus outlines estimations for orders of run times. As we see the total run time for four variables is fairly large and costly in computation time, while two variables are not likely to give us enough variation in the variables in the models, thus this paper uses three variables to train against.

While we can apply linear regression to the data, there are often complex interrelationships that need to be analysed with machine learning. In the evaluation, the data was analysed using the Python RandomForestRegressor method, using a success rate of ±7.5% for asthma hospitalisation rates. In each case, 30% of the data is taken to train the machine model, and then all of the data is used to test for success.

Table I outlines patients hospitalised with asthma for local authority areas within Scotland. In terms of correlation with hospitalised due to asthma, Table IV provides the strong correlation factors which were greater than a magnitude of 0.5 from within the dataset. This shows that COPD, smoking, and metrics related to those hospitalised over 65 were the strongest in correlation. There are thus generally strong positive correlations with COPD, but a negative correlation between male and female life expectancy.

In Table V, we see the results of running random forest linear regression against patients hospitalised with asthma against three other metrics. The determination of success in predicting the hospitalisation rate is defined as ±7.5%. The best success level is 86.67%, and which had eight best solutions (Table VI shows one of the best models). One of these is New cancer registrations, People aged 65+ with high care needs cared at home, and Children in Poverty. If we now analyse the Top 100 models from machine learning, we generate Table III. In this way, we see that Patients with emergency hospitalisation appear within 71.7% of the top models, and is a strong predictor for estimating hospitialisation rates for asthma.

TABLE I
PATIENTS HOSPITALISED WITH ASTHMA (PER 100,000 POPULATION. YEARLY AVERAGES FOR THREE YEAR PERIOD FROM 2014/2015 TO 2016/2017.)

Area	Patients hospitalised with asthma
Aberdeen City	73.6
Aberdeenshire	57.7
Angus	67.7
Argyll and Bute	71.8
Clackmannanshire	77.8
Dumfries and Galloway	80.3
Dundee City	82.9
East Ayrshire	109
East Dunbartonshire	78.4
East Renfrewshire	80.7
Edinburgh	86.8
Falkirk	91
Fife	77.4
Glasgow City	119.5
Highland	92.2
Inverclyde	103.5
Lanarkshire	109.6
Mid and East Lothian	110.3
Moray	60.2
North Ayrshire	133.1
Orkney Islands	56.6
Outer Hebrides	68.6
Perth and Kinross	63.3
Renfrewshire	104.3
Scottish Borders	86.1
Shetland Islands	39
South Ayrshire	100.9
Stirling	66
West Dunbartonshire	115.8
West Lothian	87.6

V. CONCLUSIONS

Increasingly open source data can be used to make predictions based on populations, and this paper shows that key metrics in predicting asthma rates, such as COPD, smoking rates and emergency hospitalisation rates. Within machine learning, a key factor is defining the required training features within a model. This paper has outlined some of the key features which could be used to predict asthma hospitalisation rates from open source data, and the ones which can be ignored.

VI. DATA SET

The data set is available at https://asecuritysite.com/log/well.csv and machine learning models at https://asecuritysite.com/bigdata/ml?file=well.csv.

REFERENCES

[1] L. Rutherford, S. Hinchliffe, and C. Sharp, "The Scottish Health Survey 2012," *The Scottish Health Survey*, vol. 1, 2013.

[2] E. D. Bateman, S. S. Hurd, P. J. Barnes, J. Bousquet, J. M. Drazen, M. FitzGeralde, P. Gibson, K. Ohta, P. O'Byrne, S. E. Pedersen, E. Pizzichini, S. D. Sullivanee, S. E. Wenzel, and H. J. Zar, "Global strategy for asthma management and prevention: GINA executive summary," *European Respiratory Journal*, vol. 31, no. 1, pp. 143–178, 2008.

[3] A. UK, "Connected asthma : how technology will transform care," p. 25, 2016.

[4] Statista, "Number of individuals with asthma in Europe in 2015 , by county (in million)," Tech. Rep., 2015.

[5] Asthma UK, "Asthma UK Strategy 2014 – 17 Reduce risk of asthma attacks Our long-term mission is to :," Tech. Rep., 2014.

[6] A. Sheikh, M. F. Steiner, G. Cezard, N. Bansal, C. Fischbacher, C. R. Simpson, A. Douglas, and R. Bhopal, "Ethnic variations in asthma hospital admission, readmission and death: A retrospective, national cohort study of 4.62 million people in Scotland," *BMC Medicine*, vol. 14, no. 1, pp. 1–9, 2016.

[7] Scottish Government Health Department, "Scottish Index of Multiple Deprivation," pp. –, 2017. [Online]. Available: http://www.scotland.gov.uk/Topics/Statistics/SIMD

[8] "Home - ScotPho." [Online]. Available: http://www.scotpho.org.uk/

[9] M. Martel, E. Rey, J. L. Malo, S. Perreault, M. F. Beauchesne, A. Forget, and L. Blais, "Determinants of the incidence of childhood asthma: A two-stage case-control study," *American Journal of Epidemiology*, vol. 169, no. 2, pp. 195–205, 2009.

[10] J. B. Austin, S. Selvaraj, D. Godden, and G. Russell, "Deprivation, smoking, and quality of life in asthma," *Archives of Disease in Childhood*, vol. 90, no. 3, pp. 253–257, 2005.

[11] L. Forbes, "Asthma and atopy: Endocrine or metabolic conditions?" *Thorax*, vol. 60, no. 10, pp. 793–794, 2005.

[12] J. P. Watson, P. Cowen, and R. A. Lewis, "The relationship between asthma admission rates, routes of admission, and socioeconomic deprivation," *European Respiratory Journal*, vol. 9, no. 10, pp. 2087–2093, 1996.

[13] N. Halfon and P. W. Newacheck, "Childhood asthma and poverty: differential impacts and utilization of health services," *Pediatrics*, vol. 91, no. 1, pp. 56–61, 1993.

[14] C. L. Joseph, S. L. Havstad, D. R. Ownby, C. C. Johnson, and B. C. Tilley, "Racial differences in emergency department use persist despite allergist visits and prescriptions filled for antiinflammatory medications," *Journal of Allergy and Clinical Immunology*, vol. 101, no. 4 I, pp. 484–490, 1998.

[15] R. J. Rona, "Asthma and poverty," *Thorax*, vol. 55, no. 3, pp. 239–244, 2000.

[16] S. S. Braman, "Asthma in the Elderly," *Clinics in Geriatric Medicine*, vol. 33, no. 4, pp. 523–537, 2017.

[17] F. H. Muller, "Tabakmißbrauch und Lungencarcinom," *Zeitschrift für Krebsforschung*, vol. 49, no. 1, pp. 57–85, 1940.

[18] E. Schairer and E. Schöniger, "Lungenkrebs und Tabakverbrauch," *Zeitschrift für Krebsforschung*, vol. 54, no. 4, pp. 261–269, 1944.

[19] G. E. A. Wydner EL, "Tobacco smoking as a possible etiologic factor in bronchogenic carcinoma. a study of six hundred and eighty four proved cases," *J Am Med Assoc.*, vol. 143, no. 143, pp. 329–336, 1950.

[20] R. Doll and A. B. Hill, "Lung Cancer And Other Causes Of Death In Relation To Smoking A Second Report On The Mortality Of British Doctors," *British Medical Journal*, vol. 2, no. 5001, p. 1071, 1956.

[21] S. Gompertz, D. L. Bayley, S. L. Hill, and R. A. Stockley, "Relationship between airway inflammation and the frequency of exacerbations in patients with smoking related COPD," *Thorax*, vol. 56, no. 1, pp. 36–41, 2001.

[22] K. Lasser, J. W. Boyd, S. Woolhandler, D. U. Himmelstein, D. McCormick, and D. H. Bor, "Smoking and mental illness," *Australasian psychiatry : bulletin of Royal Australian and New Zealand College of Psychiatrists*, vol. 23, no. 1, p. 2606, 2015.

[23] Y. Liu, K. Rao, T. w. Hu, Q. Sun, and Z. Mao, "Cigarette smoking and poverty in China," *Social Science and Medicine*, vol. 63, no. 11, pp. 2784–2790, 2006.

[24] K. Humphreys, "Why the wealthy stopped smoking, but the poor didn't," 2015.

[25] B. Prasadl, P. E. S. N. Krishna Prasad, and y. Sagar, "An Approach to Develop Expert Systems in Medical Diagnosis Using Machine Learning Algorithms (Asthma) and A Performance Study," *International Journal on Soft Computing*, vol. 2, no. 1, pp. 26–33, 2011. [Online]. Available: http://www.airccse.org/journal/ijsc/papers/2111ijsc03.pdf

[26] R. Howard, M. Rattray, M. Prosperi, and A. Custovic, "Distinguishing Asthma Phenotypes Using Machine Learning Approaches," *Current Allergy and Asthma Reports*, vol. 15, no. 7, p. 38, 7 2015.

[27] J. Henderson, R. Granell, J. Heron, A. Sheriff, A. Simpson, A. Woodcock, D. P. Strachan, S. O. Shaheen, and J. A. Sterne, "Associations of wheezing phenotypes in the first 6 years of life with atopy, lung function and airway responsiveness in mid-childhood," *Thorax*, vol. 63, no. 11, pp. 974–980, 2008. [Online]. Available: http://thorax.bmj.com/

[28] J. Sunyer, J. M. Antó, J. Harris, M. Torrent, O. Vall, P. Cullinan, and A. Newman-Taylor, "Maternal atopy and parity," *Clinical and Experimental Allergy*, vol. 31, no. 9, pp. 1352–1355, 2001.

[29] P. Rzehak, A. H. Wijga, T. Keil, E. Eller, C. Bindslev-Jensen, H. A. Smit, J. Weyler, S. Dom, J. Sunyer, M. Mendez, M. Torrent, O. Vall, C. P. Bauer, D. Berdel, B. Schaaf, C. M. Chen, A. Bergström, M. P. Fantini, M. Mommers, U. Wahn, S. Lau, and J. Heinrich, "Body mass index trajectory classes and incident asthma in childhood: Results from 8 European Birth Cohorts - A Global Allergy and Asthma European Network initiative," *Journal of Allergy and Clinical Immunology*, vol. 131, no. 6, pp. 1528–1536, 6 2013.

[30] G. B. Marks, S. Mihrshahi, A. S. Kemp, E. R. Tovey, K. Webb, C. Almqvist, R. D. Ampon, D. Crisafulli, E. G. Belousova, C. M. Mellis, J. K. Peat, and S. R. Leeder, "Prevention of asthma during the first 5 years of life: A randomized controlled trial," *Journal of Allergy and Clinical Immunology*, vol. 118, no. 1, pp. 53–61, 7 2006. [Online]. Available: http://www.ncbi.nlm.nih.gov/pubmed/16815138

[31] F. P. Perera, V. Rauh, W. Y. Tsai, P. Kinney, D. Camann, D. Barr, T. Bernert, R. Garfinkel, Y. H. Tu, D. Diaz, J. Dietrich, and R. M. Whyatt, "Effects of transplacental exposure to environmental pollutants on birth outcomes in a multiethnic population," *Environmental Health Perspectives*, vol. 111, no. 2, pp. 201–205, 2003. [Online]. Available: https://www.ncbi.nlm.nih.gov/pmc/articles/PMC1241351/pdf/ehp0111-000201.pdf

[32] A. Host, S. Halken, H. P. Jacobsen, A. E. Christensen, A. M. Herskind, and K. Plesner, "Clinical course of cow's milk protein allergy/intolerance and atopic diseases in childhood," *Pediatric Allergy and Immunology*, vol. 13, no. s15, pp. 23–28, 12 2002. [Online]. Available: http://doi.wiley.com/10.1034/j.1399-3038.13.s.15.7.x

[33] S. H. Downs, "Continued increase in the prevalence of asthma and atopy," *Archives of Disease in Childhood*, vol. 84, no. 1, pp. 20–23, 2001.

[34] F. Kauffmann and M.-H. Dizier, "EGEA (Epidemiological study on the Genetics and Environment of Asthma, bronchial hyperresponsiveness and atopy) - design issues," *Clinical and Experimental Allergy*, vol. 25, no. s2, pp. 19–22, 11 1995.

[35] P. Rzehak, S. Sausenthaler, S. Koletzko, C. P. Bauer, B. Schaaf, A. Von Berg, D. Berdel, M. Borte, O. Herbarth, U. Krämer, N. Fenske, H. E. Wichmann, and J. Heinrich, "Period-specific growth, overweight and modification by breastfeeding in the GINI and LISA birth cohorts up to age 6 years," pp. 449–467, 2009. [Online]. Available: http://www.jstor.org/stable/40284152

[36] A. Simpson, V. Y. Tan, J. Winn, M. Svensén, C. M. Bishop, D. E. Heckerman, I. Buchan, and A. Custovic, "Beyond atopy: Multiple patterns of sensitization in relation to asthma in a birth cohort study," *American Journal of Respiratory and Critical Care Medicine*, vol. 181, no. 11, pp. 1200–1206, 6 2010. [Online]. Available: http://www.atsjournals.org/doi/abs/10.1164/rccm.200907-1101OC

TABLE II
ESTIMATIONS OF RUN TIME FOR A RANGE OF VARIABLES TO TRAIN AGAINST

Variables to match	Combinations	Seconds per variable	Minutes per variable	Hours per variable	Total time (hours)
2	1485	594	9.9	0.165	9.24
3	26,235	10,494	174.9	2.915	163.24
4	341,055	136422	2,273.7	37.895	2,122.12
5	3478,761	1,391,504.4	23,191.74	386.529	21,645.624
6	28,989,675	11,595,870	193,264.5	3,221.075	180,380.2
7	202,927,725	81,171,090	1,352,851.5	22,547.525	1262661.4

TABLE III
ESTIMATIONS OF RUN TIME FOR A RANGE OF VARIABLES TO TRAIN AGAINST

Metric	Occurrence in machine learning model (%)
Patients with emergency hospitalisations	71.7
Patients (65+) with multiple emergency hospitalisations	26
Population prescribed drugs for anxiety/depression/psychosis	20.9
Breast screening uptake	14.9
Children Living in Poverty	13.7
Working age population claiming Out of Work benefits	10.2
Domestic Abuse	9
Working age population employment deprived	8.1
Crime rate	6.8
Adults rating neighbourhood as a very good place to live	6.8
Adults incapacity benefit/severe disability allow/employment allow	6.8
Road traffic accident casualties	6
Child obesity in primary	6
New cancer registrations	5.5
People aged 65+ with high care needs cared at home	5.1
Teenage pregnancies	4.7
Patients with a psychiatric hospitalisation	4.3
All mortality among 15-44 year olds	4.3
Deaths from suicide	4.3
Active travel to work	3.8
Child dental health in primary 1	3.8
Secondary school attendance	3.8
Immunisation uptake at 24 months-5 in 1	3
Mothers smoking during pregnancy	3
Violent crimes recorded	3
Population income deprived	3
Immunisation uptake at 24 months-MMR	2.6
Average tariff score of all pupils on S4 roll	2.6
Primary school attendance	2.6
Working age adults with low/no educational qual	2.6
Patients hospitalised with (COPD)	2.6
Early deaths from cancer (<75)	2.1
Young people not in employment education/training	2.1
Smoking prevalence (adults 16+)	2.1
Drug-related hospital stays	2.1
Prisoner population	1.7
Low birth weight	1.7
Single adult dwellings	1.7
Child dental health in primary 7	1.7
Female life expectancy	1.7
Alcohol-related hospital stays	1.3
Children looked after by local authority	1.3
People living in 15% most access deprived areas	1.3
Bowel screening uptake	1.3
Population within 500 metres of a derelict site	1.3
Patients hospitalised with coronary heart disease	1.3
Deaths all ages	0.9
Referrals Childrens Reporter-violence-related off	0.9
Pop growth (2005-2015)	0.9
Drug crimes recorded	0.4
Babies exclusively breastfed at 6-8 weeks	0.4
Estimated smoking attributable deaths	0.4
Male life expectancy	0
Early deaths from CHD (<75)	0
People claiming pension credits (aged 60+)	0

TABLE IV
PEARSON CORRELATION COEFFICIENT RELATED TO HOSPITALISED WITH ASTHMA

Pos	Metric	Correlation
1	Patients hospitalised with COPD	0.859453
2	Patients (65+) with multiple emergency hospital...	0.822299
3	Population 65+ years at risk of hospital admiss...	0.79177
4	Multiple admission patients 65+	0.790819
5	Patients with emergency hospitalisations	0.786936
6	Adults incapacity benefit/severe disability all...	0.786007
7	Estimated smoking attributable deaths	0.740913
8	Population prescribed drugs for anxiety/depress...	0.733842
9	Patients hospitalised with (COPD)	0.693275
10	Adults 65+ years claiming Attendance Allowance	0.683372
11	Children looked after by local authority	-0.683088
12	Deaths all ages	0.671886
13	People Claiming Pension Credits (aged 60+ years)	0.658914
14	Patients registered with cancer - Females	0.653211
15	Alcohol-related hospital stays	0.64417
16	Adults 60+ years claiming incapacity/severe dis...	0.62582
17	Cerebrovascular disease patients	0.624988
18	Early deaths from CHD (<75)	0.622994
19	Male life expectancy	-0.622139
20	Working age adults with low/no educational qual	0.621369
21	Single adult dwellings	0.619495
22	Drug-related hospital stays	0.615702
23	Early deaths from cancer (<75)	0.614767
24	New cancer registrations	0.602819
25	Population (65+) in 15% most access deprived areas	-0.597966
26	Patients hospitalised with coronary heart disease	0.590585
27	Percent of low birthweight (less than 2500g) ba...	0.553096
28	Female life expectancy	-0.547162
29	Deaths from alcohol conditions	-0.524465

TABLE V
MACHINE LEARNING METRICS FOR PATIENTS HOSPITALISED WITH ASTHMA WITH A 86.67% SUCCESS RATE OF PREDICTION

New cancer registrations	People aged 65+ with high care needs cared at home	Children Living in Poverty
Patients (65+) with multiple emergency hospitalisations	Deaths from suicide	Breast screening uptake
Patients (65+) with multiple emergency hospitalisations	Children looked after by local authority	Breast screening uptake
Patients (65+) with multiple emergency hospitalisations	Single adult dwellings	Breast screening uptake
Patients (65+) with multiple emergency hospitalisations	Secondary school attendance	Breast screening uptake
Patients (65+) with multiple emergency hospitalisations	Children Living in Poverty	Adults rating neighbourhood as a very good place to live
Patients (65+) with multiple emergency hospitalisations	Children Living in Poverty	Breast screening uptake
Population prescribed drugs for anxiety/depression/psychosis	People aged 65+ with high care needs cared at home	Working age adults with low/no educ qual

TABLE VI
TRAINING DATA: NEW CANCER REGISTRATIONS, PEOPLE AGED 65+ WITH HIGH CARE NEEDS CARED AT HOME AND CHILDREN LIVING IN POVERTY. TRAINED AGAINST: PATIENTS HOSPITALISED WITH ASTHMA WITH 30% TRAINING DATA

Index	Predicted	Actual	Diff	Success
Aberdeen City	73.64	73.60	0.04	Success
Aberdeenshire	56.67	57.70	-1.03	Success
Angus	65.69	67.70	-2.01	Success
Argyll and Bute	75.11	71.80	3.31	Success
Clackmannanshire	83.33	77.80	5.53	Success
Dumfries and Galloway	83.83	80.30	3.53	Success
Dundee City	85.45	82.90	2.55	Success
East Ayrshire	104.56	109.00	-4.44	Success
East Dunbartonshire	75.25	78.40	-3.15	Success
East Renfrewshire	75.78	80.70	-4.92	Success
Edinburgh City	94.06	86.80	7.26	Failed!
Falkirk	90.35	91.00	-0.65	Success
Fife	80.98	77.40	3.58	Success
Glasgow City	116.11	119.50	-3.39	Success
Highland	65.61	92.20	-26.59	Failed!
Inverclyde	104.67	103.50	1.17	Success
Lanarkshire	105.53	109.60	-4.07	Success
Mid and East Lothian	103.27	110.30	-7.03	Success
Moray	61.28	60.20	1.08	Success
North Ayrshire	106.91	133.10	-26.19	Failed!
Orkney Islands	52.13	56.60	-4.47	Success
Outer Hebrides	65.66	68.60	-2.94	Success
Perth and Kinross	64.72	63.30	1.42	Success
Renfrewshire	96.24	104.30	-8.06	Failed!
Scottish Borders	69.97	86.10	-16.13	Failed!
Shetland Islands	57.29	39.00	18.29	Failed!
South Ayrshire	96.40	100.90	-4.50	Success
Stirling	68.98	66.00	2.98	Success
West Dunbartonshire	113.28	115.80	-2.52	Success
West Lothian	87.35	87.60	-0.25	Success

Quantitative Analysis of Facial Expression for Medical Doctors

Akemi Oe, Yoshitoshi Murata
Graduate School of Software and Information Science
Iwate Prefectural University
Iwate, Japan
e-mails: akemioe@gmail.com, y-murata@iwate-pu.ac.jp

***Abstract*— Establishing trust between a patient and a doctor depends as much on their relationship as on the doctor's medical abilities. An important factor in building and maintaining a relationship is whether the doctor produces facial expressions appropriate to the patient's condition. The purpose of this study is to identify facial expressions appropriate to various patient conditions. We focused on the greetings given by young doctors at the beginning of medical interviews of adult patients in the general ward of a hospital. Images of a role-playing patient portraying one of three physical conditions were shown to seven student doctors, who were then videotaped as they greeted the "patient". We identified appropriate facial expressions for each condition by comparison between human evaluation and analysis of the computer emotion analysis system.**

Keywords-doctor-patient interaction; facial expression; nonverbal communication.

I. INTRODUCTION

Patient satisfaction is an important component of medical care [1]. Improving patient satisfaction enhances trust and the relationship between patient and doctor, which leads to stronger adherence to the prescribed protocol, such as taking medicine, and to enhanced therapeutic effect [2][3]. Many studies and reviews have shown that the main determinant of patient satisfaction is patient-doctor relationship [4]-[8] and that patient satisfaction is higher when the patient communicates with a doctor having strong nonverbal communication ability [9][10]. However, inexperienced young doctors and medical students often have trouble producing appropriate facial expressions when greeting a patient. The first author of this paper, a lecturer on medical communication, often hears young doctors complaining that, though they intend to smile, patients say that they seem to be angry.

In this study, we identified appropriate facial expressions for a doctor by using quantitative analysis. We videotaped medical students' facial expressions when they greeted a patient and analyzed the recorded videos using the computer emotion analysis system. We then asked potential patients to evaluate the appropriateness of the facial expressions on a five-point scale. We clarify a facial expression found according to the scene by a doctor by comparing both.

The facial expression required for a doctor may be influenced by the patient's condition, treatment scene, culture, medical department, gender. As the first step, we focused on appropriate facial expressions of young medical doctors for greeting patients in the general ward of a hospital.

After reviewing related works in Section II, our facial expression analysis system is explained in Section III. We describe the evaluation experiment in Section IV, and analyze facial expression for recorded videos in Section V. We conclude with a summary of the key points and a mention of future work in Section VI.

II. RELATED WORK

A. Medical Interviews

Medical interviews have traditionally focused on gathering relevant information from patients [11]. Nowadays, the focus has expanded to building a trusting relationship, sharing decision-making, responding to the patient's emotional state, supporting actions related to the patient's condition and treatment, so the doctor must have a wider range of communication skills [12]. These skills include "looking at a patient not as a case but as a human being" [13] and "building and maintaining a good relationship between doctor and patient" [14]. It has been shown that such skills have a greater effect on patient satisfaction than the doctor's medical skills, the medicine prescribed, the information provided, the questions asked, the advice given, and the instructions given. In particular, a patient's satisfaction is positively related to the doctor being warm [13][15], empathic [13][15]-[17], and friendly [15] and giving the impression of being human [16].

"Nonverbal communication" is a means of communicating these emotional aspects. Patient satisfaction is higher when the doctor has a strong ability to express his or her emotions and to read the emotions of others by nonverbal communication such as through facial expressions, gaze, posture, and tone of voice [9][18][19]. In short, a doctor's nonverbal communication is an important aspect of patient care.

B. Nonverbal Communication

Facial expression plays a large role in nonverbal communication. For emotional messages such as "likes" and "dislikes," Mehrabian [20] estimated that 7% of the message is carried by the language content, 38% is carried by the voice and sound quality, and 55% is carried by the facial expression and gestures. Birdwhistell [21] argued that 35% of the message is carried by the language content while the remaining 65% is carried by the expression, the way of talking, the gestures, etc. Therefore, it is useful to clarify the appropriate facial expressions for doctors to have when communicating with patients, especially patients who are sensitive to a doctor's nonverbal behavior due to anxiety [22] [23].

III. FACIAL EXPRESSION ANALYSIS SYSTEM

To identify appropriate facial expressions, we developed a system that quantitatively analyzes changes in facial expression. It is based on the Cognitive Services Emotion API [24] provided by Microsoft's Azure cloud service. The concept of the system is illustrated in Figure 1. The doctor's facial image during a patient interview is recorded in a video file. The file is sent to the Cognitive Services Emotion API, which provides feedback on the position coordinates of the doctor's face and the ratio by emotion. To realize the above functions, the developed system comprises Microsoft.ProjectOxford.Emotion.dll which corresponds to the Emotion API on Azure, Newtonsoft.Json.dll which handles the file in JSON format, Parakeet.dll which processes the movie file, and Parakeet.Logging.dll which passes the facial expression analysis received from the Emotion API as a log file to for real time display. We developed Parakeet.dll, Parakeet.Logging.dll and LogViewerWPF.exe.

Our facial expression emotion analysis system calculates the ratio for seven emotions ("happiness," "anger," "contempt," "disgust," "fear," "sadness," and "surprise") reflected in the input video image, and "neutral." The total for all emotions is 1, and the value for "neutral" is obtained by subtracting the total value for the seven emotions from 1.

The detection result window is shown in Figure 2. Each row shows the results for the emotion corresponding to one facial expression. The rows are in time series order, with the latest set of results in the bottom row. Clicking the display command on the menu highlights the detection results for the specified face. Selecting a line by using the mouse or keyboard causes the corresponding video to play.

The cells corresponding to each emotional value are highlighted in "pink." If the value = 1, the cells are the darkest pink, and if the value = 0, the cells have no color. For example, if the emotional value is 0.8, the color density is 80%.

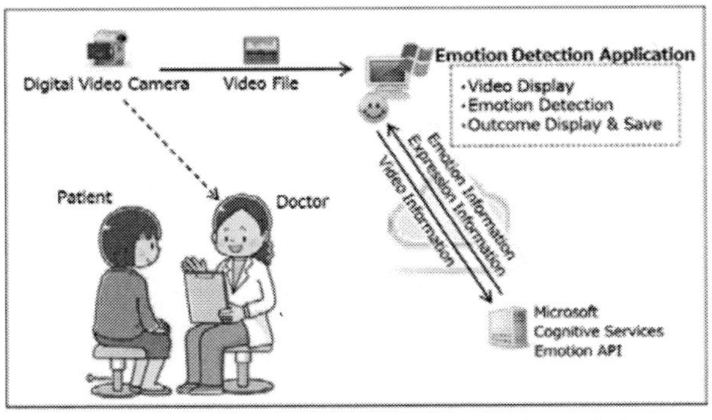

Figure 1. Configuration of video image-based emotion analysis system.

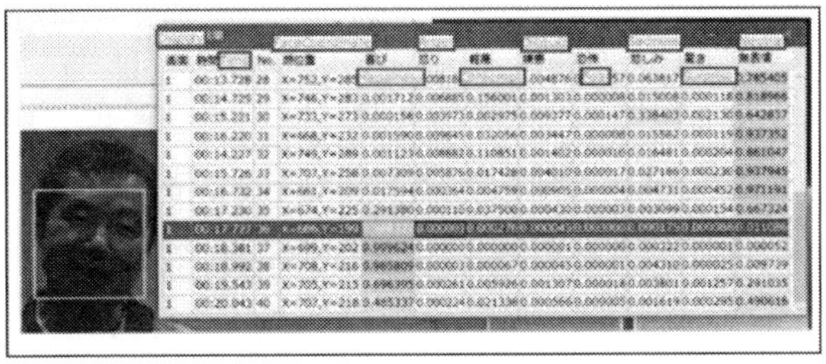

Figure 2. Detection result window.

IV. EVALUATION EXPERIMENT

One way to analyze the facial expressions of veteran doctors would be to clarify facial expressions appropriate for doctors. However, their facial expression would not be always right. And, since many young physicians have trouble presenting appropriate facial expressions when greeting a patient, we chose to find facial expressions that would be acceptable for most patients including potential patients from facial expressions which medical students think suitable. This time, we targeted the situation when they greet patients in the general ward of a hospital. We videotaped their greeting for model patients to evaluate and identified appropriate facial expressions. Although evaluation by actual patients is best, it would have been difficult to request their participation. Hence, we asked general healthy adults who had hospitalization in the past or would have ones in the future. At the same time, we used computerized facial expression analysis to identify suitable facial expressions, as introduced in section III.

A. *Experimental Conditions*

1) Doctor participants: Seven medical students (4 females, 3 males; average age 22.5 years)

2) Patient conditions: Although the actual condition of hospitalized patients varies widely among patients, we had a role-playing patient portray only three conditions, as shown in Figure 3: a patient who feels physically healthy (a "bright patient"), one whose physical condition is unknown (an "expressionless patient"), and one who feels badly and is suffering pain (a "patient in pain").

3) Video recording: The three photographs in Figure 3 were presented to the doctor participants along with the following explanation. "The pictures you will see show a patient you visit during regular morning rounds in a hospital. They are of the same patient, but his condition is different in each picture. In the first one, he feels physically healthy, in the second one, his physical condition is unknown, and in the third one, he feels badly and is suffering pain. After looking at each picture, please greet the patient as a doctor for about 5 seconds or so." This process was repeated days later using three of the students and the same photographs. On the first day, all students greeted the patient as they thought best. Before the process on the second day, the three selected students received coaching based on the authors'

Figure 3. Conditions portrayed by role-playing patient.

previous findings. On the second day, foregoing three students greeted the patient again. We recorded their greetings (30 recordings in total) and used them for our evaluation of facial expressions.

The purpose of this research is not to find most appropriate facial expressions, but acceptable ones for most patients. Therefore, we think not many doctors are needed as participants. In this paper, we consider the meaning of appropriate to be the same as acceptable.

B. Subjective Evaluation

To make the subjective evaluation more effective, we first had 16 people view and evaluate each video recording and removed the ones in which the student's facial expression was judged to be unacceptable. We then had 31 other people view and evaluate the remaining recordings.

1) First subjective evaluation: We showed the video recordings to 5 men and 11 women (average age 46) without sound. We asked them to judge whether the doctor's facial expression was appropriate for the situation on a 5-point scale ("1 completely appropriate," "2 somewhat appropriate," "3 neutral," "4 not so appropriate," "5 not at all appropriate "). We asked them to also comment on anything they felt or noticed. We showed the recordings without sound because we wanted them to focus on appropriate facial expressions in medical communication situations, and emotion is easier to read from speech than from facial expressions. The results are shown in Table I.

2) Second subjective evaluation: For the second subjective evaluation, we eliminated the recordings with score of 4 or 5, these would be unacceptable values for patients, in the first evaluation, except for one score-4 recording. The score-4 recording was kept because otherwise there would have been only three recordings for the "patient in pain." The videos used were B-1 to B-6 for the "bright patient" (shown in blue in Table I), E-1, E-2, E-3, E-4, and E-8 for the "expressionless patient" (shown in gray), and P-1 to P-4 and P-8 for the "patient in pain" (shown in yellow).

We showed these videos to 17 men and 14 women (average age 35.9) without sound. We asked them to judge whether the doctor's facial expression was appropriate for the patient's condition on the same 5-point scale. We again asked them to also comment on anything they felt or noticed. The results are shown in Table II. As in the first evaluation, we showed the videos without sound so that they would focus their attention on the facial expressions.

TABLE I. RESULTS OF FIRST SUBJECTIVE EVALUATION

Video-ID		F51	F67	F65	M54	F59	F58	F57	M48	F55	F50	F20	F27	F24	M34	M34	M32
Bright	B-1	1			2			1	2	1	1	2	2	1	1	2	1
	B-2	1	2	2	2	1	1	1	2	1	1	2	1	1	2	2	1
	B-3	1			2			2	2	1	1	2	2	1	2	1	1
	B-4	1	2	1	2	1	1	2	1	2	2	1	2	2	2	2	1
	B-5							1	1	2	1	2	2	1	2	2	2
	B-6							2	2	2	1	2	2	2	2	2	1
	B-7	1			4			3	1	1	1	2	3	2	2	2	1
	B-8	1	2	1	2	1	2	3	2	2	3	2	2	2	4	4	2
	B-9	1	1	2	4	3	2	2	4	2	4	2	2	3	3	2	3
	B-10	2			2			2	4	1	2	2	3	3	4	3	4
Expressionless	E-1	1			2			3	1	2	1	3	3	1	2	1	2
	E-2	1			2			2	2	3	1	2	3	2	2	2	2
	E-3							2	3	2	2	2	2	2	2	2	1
	E-4							2	3	2	2	2	2	2	2	2	1
	E-5	1			2			3	2	1	1	3	4	2	2	2	2
	E-6	1	2	2	2	4	2	2	2	4	2	2	2	2	2	2	1
	E-7	1	2	3	2	1	2	1	4	3	4	3	1	3	3	2	2
	E-8	2	2	1	2	3	1	2	2	2	3	3	2	3	3	3	3
	E-9	2	3	2	3	4	3	2	1	3	2	3	2	3	3	4	1
	E-10	2			2			3	4	1	2	3	3	3	4	2	4
in Pain	P-1							2	2	1	2	2	2	3	2	2	1
	P-2	1			2			2	3	1	1	3	1	3	2	3	2
	P-3							1	1	1	2	2	2	3	3	3	2
	P-4	2			2			1	1	2	2	2	4	2	2	2	3
	P-5	2	2	2	2	2	2	2	3	3	2	1	3	4	2	2	4
	P-6	2	2	2	2	2	2	2	3	3	2	1	3	4	2	2	4
	P-7	2	2	2	4	2	3	2	2	3	2	1	2	4	4	3	3
	P-8	2			2			3	2	3	2	3	3	3	4	2	2
	P-9	1	2	3	4	2	3	4	3	3	3	4	2	5	4	4	1
	P-10	2			4			4	4	2	2	4	3	3	2	3	4

TABLE II. RESULTS OF COMPUTER EVALUATION

Video-ID		Score					avg
		1	2	3	4	5	
Bright	B-2	9	17	4	1	0	1.9
	B-4	13	8	9	1	0	1.9
	B-3	7	10	12	2	0	2.3
	B-5	2	15	13	1	0	2.4
	B-6	5	8	16	2	0	2.5
	B-1	3	9	16	3	0	2.6
Expressionless	E-1	4	10	14	3	0	2.5
	E-3	5	5	19	2	0	2.6
	E-4	4	6	18	3	0	2.6
	E-8	3	5	22	0	1	2.7
	E-2	1	6	13	11	0	3.1
in Pain	P-4	10	8	9	4	0	2.2
	P-3	5	8	16	2	0	2.5
	P-1	3	9	13	5	1	2.7
	P-8	2	10	8	10	1	2.9
	P-2	0	4	14	11	2	3.4

V. QUANTITATIVE ANALYSIS

In addition to the two subjective evaluations, the emotions represented by the facial expressions in the 16 highest ranked videos were identified using the emotion analysis system.

A. *Facial expressions appropriate for "bright patient"*

1) The three top-ranked videos for "bright patient" as subjectively evaluated were analyzed from the point of view of what emotions appear in the facial expressions, as identified by computer. Tables III, IV, and V show the results. Table VI shows the results for video B-10, which was evaluated as low. The cells in the tables corresponding to 0 or more and less than 0.2 are shown in blue, 0.2 or more and less than 0.4 in green, 0.4 or more and less than 0.6 in yellow, 0.6 or more and less than 0.8 in orange, and 0.8 or more in red. A representative facial expression is shown in Figure 4. Images of the students corresponding to videos B-2 and B-4 are not shown because we could not get permission.

Figure 4. Facial expression of student corresponding to video B-3.

TABLE III. COMPUTER ANALYSIS RESULT FOR VIDEO B-2 (TOP-RANKED FACIAL EXPRESSION FOR "BRIGHT PATIENT")

Time	Happiness	Anger	Contempt	Disgust	Fear	Sadness	Surprise	Neutral
00:00.1	0.96	0.00	0.00	0.00	0.00	0.00	0.00	0.04
00:00.7	0.95	0.00	0.00	0.00	0.00	0.00	0.00	0.05
00:01.1	0.90	0.00	0.00	0.00	0.00	0.00	0.00	0.10
00:01.6	1.00	0.00	0.00	0.00	0.00	0.00	0.00	0.00
00:02.1	1.00	0.00	0.00	0.00	0.00	0.00	0.00	0.00
00:02.8	1.00	0.00	0.00	0.00	0.00	0.00	0.00	0.00
00:03.4	1.00	0.00	0.00	0.00	0.00	0.00	0.00	0.00
00:03.9	1.00	0.00	0.00	0.00	0.00	0.00	0.00	0.00
00:04.4	1.00	0.00	0.00	0.00	0.00	0.00	0.00	0.00

TABLE IV. COMPUTER ANALYSIS RESULT FOR VIDEO B-4 (2ND-RANKED FACIAL EXPRESSION FOR "BRIGHT PATIENT")

Time	Happiness	Anger	Contempt	Disgust	Fear	Sadness	Surprise	Neutral
00:00.1	0.83	0.00	0.01	0.01	0.00	0.00	0.00	0.16
00:00.7	0.99	0.00	0.00	0.00	0.00	0.00	0.00	0.01
00:01.1	0.98	0.00	0.00	0.01	0.00	0.00	0.00	0.01
00:01.7	1.00	0.00	0.00	0.00	0.00	0.00	0.00	0.00
00:02.1	1.00	0.00	0.00	0.00	0.00	0.00	0.00	0.00
00:02.8	1.00	0.00	0.00	0.00	0.00	0.00	0.00	0.00
00:03.2	0.97	0.00	0.00	0.02	0.00	0.00	0.01	0.00
00:03.7	0.95	0.00	0.00	0.01	0.00	0.00	0.00	0.04
00:04.2	1.00	0.00	0.00	0.00	0.00	0.00	0.00	0.00
00:04.8	0.99	0.00	0.00	0.01	0.00	0.00	0.00	0.00
00:05.4	0.99	0.00	0.00	0.00	0.00	0.00	0.00	0.00
00:06.0	1.00	0.00	0.00	0.00	0.00	0.00	0.00	0.00
00:06.5	1.00	0.00	0.00	0.00	0.00	0.00	0.00	0.00
00:06.9	1.00	0.00	0.00	0.00	0.00	0.00	0.00	0.00

TABLE V. COMPUTER ANALYSIS RESULT FOR VIDEO B-3 (3RD-RANKED FACIAL EXPRESSION FOR "BRIGHT PATIENT")

Time	Happiness	Anger	Contempt	Disgust	Fear	Sadness	Surprise	Neutral
00:00.1	1.00	0.00	0.00	0.00	0.00	0.00	0.00	0.00
00:00.7	1.00	0.00	0.00	0.00	0.00	0.00	0.00	0.00
00:01.1	1.00	0.00	0.00	0.00	0.00	0.00	0.00	0.00
00:02.2	1.00	0.00	0.00	0.00	0.00	0.00	0.00	0.00
00:01.7	1.00	0.00	0.00	0.00	0.00	0.00	0.00	0.00
00:02.8	1.00	0.00	0.00	0.00	0.00	0.00	0.00	0.00
00:03.3	1.00	0.00	0.00	0.00	0.00	0.00	0.00	0.00
00:03.9	1.00	0.00	0.00	0.00	0.00	0.00	0.00	0.00

TABLE VI. COMPUTER ANALYSIS RESULT FOR VIDEO B-10 (LOW-RANKED FACIAL EXPRESSION FOR "BRIGHT PATIENT")

Time	Happiness	Anger	Contempt	Disgust	Fear	Sadness	Surprise	Neutral
00:00.0	0.00	0.00	0.00	0.00	0.00	0.00	0.00	1.00
00:00.6	0.00	0.00	0.00	0.00	0.03	0.02	0.50	0.45
00:01.1	0.00	0.00	0.00	0.02	0.00	0.03	0.00	0.97
00:01.6	0.00	0.00	0.01	0.00	0.00	0.00	0.00	0.99
00:02.1	0.00	0.00	0.00	0.00	0.01	0.10	0.03	0.85
00:02.6	0.00	0.00	0.00	0.00	0.06	0.00	0.49	0.44
00:03.1	0.02	0.00	0.02	0.00	0.00	0.03	0.01	0.92
00:03.6	0.02	0.00	0.00	0.00	0.00	0.00	0.00	0.98
00:04.1	0.04	0.00	0.00	0.00	0.00	0.00	0.00	0.96
00:04.7	0.02	0.00	0.01	0.00	0.00	0.00	0.00	0.97

The facial expression for "bright patient" was "constant happiness" (expressed more as a laugh rather than simply a smile) for the three top-ranked videos. The lower evaluated video, B-10, was mostly "neutral," and it was judged as showing "nervousness," "no expression," "scary eye,", etc. which explains why it received a low evaluation.

B. *Facial expressions appropriate for "expressionless patient"*

The three top-ranked videos for "expressionless patient" as subjectively evaluated were analyzed from the point of view of what emotions appear in the facial expressions as identified by computer. Tables VII, VIII, and IX show the results. Table X shows the results for video E-8, which was evaluated as low.

As shown in Figure 5, the image representing the first half of video E-1 is largely "happiness" (expressed as a smile) and "neutral" in the second half (mainly expressionless). The 2nd- and 3rd-ranked were mostly "neutral." The lower evaluated video, E-8, was mostly "happiness," which was judged not to be serious enough. It is thought that this is because medical students cannot judge whether an "expressionless patient" is in a good or bad physical condition due to the lack of expression. Since the patient's condition could be bad, an expression showing "happiness" was judged by some as inappropriate. The "neutral" expressions shown in the 2nd- and 3rd-ranked videos were apparently judged as having little effect on the patient. The top-ranked video showed a natural greeting starting with a smile and then transitioning to "neutral" as the "doctor" learned about the patient's condition, which is considered to be the reason for the high evaluation.

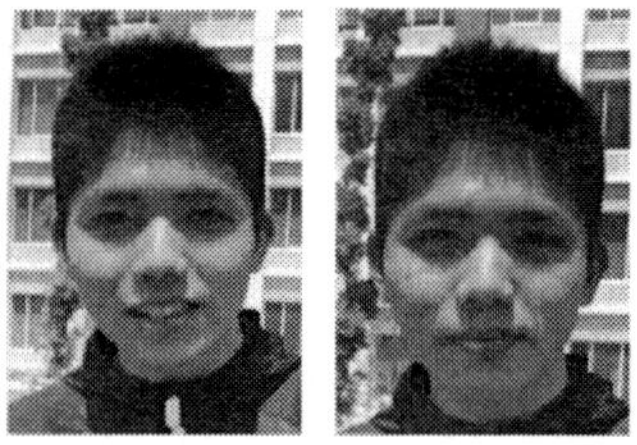

Figure 5. Facial expression of student corresponding to video E-1 (left: first half; right: second half).

TABLE VII. COMPUTER ANALYSIS RESULTS FOR VIDEO E-1 (TOP-RANKED FACIAL EXPRESSION FOR "EXPRESSIONLESS PATIENT")

Time	Happiness	Anger	Contempt	Disgust	Fear	Sadness	Surprise	Neutral
00:00.1	0.66	0.00	0.00	0.00	0.00	0.00	0.00	0.34
00:00.7	0.71	0.00	0.00	0.00	0.00	0.00	0.00	0.29
00:01.2	0.79	0.00	0.01	0.00	0.00	0.00	0.00	0.20
00:01.6	0.94	0.00	0.00	0.00	0.00	0.00	0.00	0.06
00:02.2	0.35	0.00	0.01	0.00	0.00	0.00	0.00	0.63
00:02.7	0.93	0.00	0.00	0.00	0.00	0.00	0.00	0.07
00:03.2	0.24	0.00	0.00	0.00	0.00	0.00	0.00	0.76
00:03.8	0.29	0.00	0.00	0.00	0.00	0.00	0.00	0.70
00:04.0	0.40	0.00	0.00	0.00	0.00	0.00	0.00	0.59

TABLE VIII. COMPUTER ANALYSIS RESULTS FOR VIDEO E-3 (2ND-RANKED FACIAL EXPRESSION FOR "EXPRESSIONLESS PATIENT")

Time	Happiness	Anger	Contempt	Disgust	Fear	Sadness	Surprise	Neutral
00:00.7	0.00	0.00	0.00	0.00	0.00	0.07	0.00	0.93
00:01.1	0.00	0.00	0.00	0.00	0.00	0.06	0.02	0.92
00:01.6	0.00	0.00	0.00	0.00	0.00	0.04	0.00	0.96
00:02.2	0.00	0.00	0.00	0.00	0.00	0.08	0.01	0.92
00:02.6	0.00	0.00	0.00	0.00	0.00	0.33	0.00	0.66
00:00.1	0.00	0.00	0.00	0.00	0.00	0.00	0.00	1.00
00:03.2	0.00	0.01	0.02	0.00	0.00	0.13	0.00	0.84
00:03.6	0.00	0.00	0.01	0.00	0.00	0.05	0.00	0.94
00:03.0	0.00	0.00	0.01	0.00	0.00	0.06	0.00	0.92

TABLE IX. COMPUTER ANALYSIS RESULTS FOR VIDEO E-4 (3RD-RANKED FACIAL EXPRESSION FOR "EXPRESSIONLESS PATIENT")

Time	Happiness	Anger	Contempt	Disgust	Fear	Sadness	Surprise	Neutral
00:00.1	0.02	0.00	0.00	0.00	0.00	0.00	0.01	0.96
00:01.6	0.00	0.00	0.00	0.00	0.02	0.01	0.34	0.63
00:02.1	0.00	0.00	0.00	0.00	0.03	0.01	0.35	0.60
00:02.6	0.00	0.00	0.00	0.00	0.08	0.02	0.37	0.52
00:03.1	0.00	0.00	0.00	0.00	0.05	0.05	0.24	0.65
00:03.6	0.03	0.00	0.02	0.00	0.00	0.03	0.01	0.91
00:04.1	0.00	0.00	0.01	0.00	0.00	0.03	0.00	0.96

TABLE X. COMPUTER ANALYSIS RESULTS FOR VIDEO E-8 (LOW-RANKED FACIAL EXPRESSION FOR "EXPRESSIONLESS PATIENT")

Time	Happiness	Anger	Contempt	Disgust	Fear	Sadness	Surprise	Neutral
00:00.0	0.81	0.00	0.00	0.00	0.00	0.00	0.00	0.19
00:00.5	0.60	0.00	0.01	0.00	0.00	0.00	0.05	0.34
00:01.0	0.81	0.00	0.00	0.01	0.00	0.00	0.09	0.09
00:01.6	0.96	0.00	0.00	0.00	0.00	0.00	0.00	0.03
00:02.0	0.95	0.00	0.00	0.00	0.00	0.00	0.01	0.04
00:02.5	0.96	0.00	0.00	0.01	0.00	0.00	0.00	0.03
00:03.0	0.98	0.00	0.00	0.00	0.00	0.00	0.00	0.01
00:03.7	0.91	0.00	0.02	0.01	0.00	0.00	0.00	0.07
00:04.0	0.87	0.00	0.04	0.01	0.00	0.00	0.00	0.08

C. *Facial expressions appropriate for "patient in pain"*

The three top-ranked videos for "patient in pain" as subjectively evaluated were analyzed from the point of view of what emotions appear in the facial expressions as identified by computer. Tables XI, XII, and XIII show the results. The expression in the top-ranked video is shown in Figure 6. Table XIV shows the results for video P-10, which was evaluated low.

The expressions in the higher ranked videos are mainly "neutral," with "surprise," "sadness," or "fear" gradually appearing in some. There was virtually no expression of "happiness," "anger," or "disgust" except for the third frame of video P-1. In contrast, expression in the lower ranked video, P-10, was mostly "happiness," resulting in comments such as "the grinning made me feel uncomfortable " and "the doctor seemed to be smiling faintly." These comments explain the low evaluation.

Figure 6. Facial expression of student corresponding to video P-4.

TABLE XI. COMPUTER ANALYSIS RESULTS FOR VIDEO P-4 (TOP-RANKED FACIAL EXPRESSION FOR "PATIENT IN PAIN")

Time	Happiness	Anger	Contempt	Disgust	Fear	Sadness	Surprise	Neutral
00:00.7	0.00	0.00	0.00	0.00	0.00	0.07	0.00	0.93
00:01.1	0.00	0.00	0.00	0.00	0.00	0.06	0.02	0.92
00:01.6	0.00	0.00	0.00	0.00	0.00	0.04	0.00	0.96
00:02.2	0.00	0.00	0.00	0.00	0.00	0.08	0.01	0.92
00:02.6	0.00	0.00	0.00	0.00	0.00	0.33	0.00	0.66
00:00.1	0.00	0.00	0.00	0.00	0.00	0.00	0.00	1.00
00:03.2	0.00	0.01	0.02	0.00	0.00	0.13	0.00	0.84
00:03.6	0.00	0.00	0.01	0.00	0.00	0.05	0.00	0.94
00:03.0	0.00	0.00	0.01	0.00	0.00	0.06	0.00	0.92

TABLE XII. COMPUTER ANALYSIS RESULTS FOR VIDEO P-3 (2ND-RANKED FACIAL EXPRESSION FOR "PATIENT IN PAIN")

Time	Happiness	Anger	Contempt	Disgust	Fear	Sadness	Surprise	Neutral
00:00.1	0.02	0.00	0.00	0.00	0.00	0.00	0.01	0.96
00:01.6	0.00	0.00	0.00	0.00	0.02	0.01	0.34	0.63
00:02.1	0.00	0.00	0.00	0.00	0.03	0.01	0.35	0.60
00:02.6	0.00	0.00	0.00	0.00	0.08	0.02	0.37	0.52
00:03.1	0.00	0.00	0.00	0.00	0.05	0.05	0.24	0.65
00:03.6	0.03	0.00	0.02	0.00	0.00	0.03	0.01	0.91
00:04.1	0.00	0.00	0.01	0.00	0.00	0.03	0.00	0.96

TABLE XIII. COMPUTER ANALYSIS RESULTS FOR VIDEO P-1 (3RD-RANKED FACIAL EXPRESSION FOR "PATIENT IN PAIN")

Time	Happiness	Anger	Contempt	Disgust	Fear	Sadness	Surprise	Neutral
00:00.0	0.00	0.00	0.00	0.00	0.00	0.02	0.05	0.93
00:00.6	0.01	0.00	0.00	0.00	0.00	0.00	0.06	0.93
00:01.1	0.26	0.00	0.00	0.00	0.00	0.00	0.00	0.73
00:01.6	0.02	0.00	0.00	0.00	0.00	0.00	0.00	0.98
00:02.1	0.00	0.00	0.00	0.00	0.01	0.09	0.09	0.81
00:02.6	0.05	0.00	0.00	0.00	0.00	0.00	0.00	0.94
00:03.2	0.00	0.00	0.01	0.00	0.00	0.03	0.00	0.96
00:03.6	0.00	0.00	0.00	0.00	0.00	0.00	0.00	0.99
00:04.1	0.00	0.00	0.00	0.00	0.00	0.00	0.00	1.00
00:04.6	0.00	0.00	0.00	0.00	0.00	0.00	0.00	1.00
00:05.2	0.00	0.00	0.00	0.00	0.00	0.00	0.00	1.00
00:05.0	0.01	0.00	0.00	0.00	0.00	0.00	0.00	0.98

TABLE XIV. COMPUTER ANALYSIS RESULTS FOR VIDEO P-10 (LOW-RANKED FACIAL EXPRESSION FOR "PATIENT IN PAIN")

Time	Happiness	Anger	Contempt	Disgust	Fear	Sadness	Surprise	Neutral
00:00.1	1.00	0.00	0.00	0.00	0.00	0.00	0.00	0.00
00:00.5	1.00	0.00	0.00	0.00	0.00	0.00	0.00	0.00
00:01.0	1.00	0.00	0.00	0.00	0.00	0.00	0.00	0.00
00:01.5	0.83	0.00	0.00	0.00	0.00	0.00	0.00	0.17
00:02.0	0.95	0.00	0.00	0.00	0.00	0.00	0.00	0.05
00:02.5	0.99	0.00	0.00	0.00	0.00	0.00	0.00	0.01
00:03.0	1.00	0.00	0.00	0.00	0.00	0.00	0.00	0.00
00:03.5	0.99	0.00	0.00	0.00	0.00	0.00	0.00	0.01
00:04.1	0.58	0.00	0.00	0.00	0.00	0.00	0.00	0.41
00:04.7	0.24	0.00	0.04	0.00	0.00	0.00	0.00	0.72

VI. CONCLUSION

Our quantitative analysis of medical student's facial expressions when greeting an adult patient to be medically evaluated in the general ward of a hospital revealed which facial expressions are most acceptable. For patients who feel physically healthy, the most acceptable facial expression is "continuous happiness" (expressed more as a laugh rather than simply a smile). For patients without a facial expression, the most acceptable facial expression is initially "happiness" (expressed as a smile) and then "neutral" (without expression). For patients in bad physical condition suffering pain, the most acceptable facial expression is "neutral" with a little "sadness" or "surprise."

We think appropriate facial expressions required for a medical doctor would be different for the treatment department, generation, gender and culture. We plan to clarify their influence in future work. In addition, we are developing a training system based on our findings to learn acceptable expressions.

REFERENCES

[1] A. Donabedian, The Definition of Quality and Approaches to Its Assessment. Ann Arbor, MI: Health Administration Press, 1980.

[2] S. H. Kaplan, S. Greenfield, and W. E. Ware, Jr, "Assessing the effects of physician-patient interactions on the outcomes of chronic disease," Med. Care, vol. 27, pp. S110–S127, March 1989.

[3] L. M. Noble, A. Kubacki, J. Martin, and M. Lloyd, "The effect of professional skills training on patient-centredness and confidence in communicating with patients," Med. Educ., vol. 41, pp. 432–440, May 2007.

[4] J. A. Hall, D. L. Roter, and N. R. Katz, "Meta-analysis of correlates of provider behavior in medical encounters," Med. Care, vol. 26, pp. 657–675, July 1988.

[5] J. A. Hall, J. A. Harrigan, and R. Rosenthal, "Nonverbal behavior in clinician-patient interaction," Appl. Prev. Psychol., vol. 4, pp. 21–37, Winter 1995.

[6] L. M. Ong, J. C. de Haes, A. M. Hoos, and F. B. Lammes, "Doctor-patient communication: A review of the literature," Soc. Sci. Med., vol. 40, pp. 903–918, April 1995.

[7] T. L. Thompson, "Interpersonal communication and health care," in Handbook of Interpersonal Communication, 2nd ed., M. L. Knapp and G. R. Miller, Eds. Thousand Oaks, CA: Sage, pp. 696–725, 1994.

[8] S. Williams, J., Weinman, J, and J. Dale, "Doctor-patient communication and patient satisfaction: A review," Family Practice, vol. 15, pp. 480–492, October 1998.

[9] M. R. DiMatteo, A. Taranta, H. S. Friedman, and L. M. Prince, "Predicting patient satisfaction from physicians' nonverbal communication skills," Med. Care, vol. 18, pp. 376–387, April 1980.

[10] M. R. DiMatteo, R. D. Hays, and L. M. Prince, "Relationship of physicians' nonverbal communication skill to patient satisfaction, appointment noncompliance, and physician workload," Health Psychol., vol. 5, pp. 581–594, 1986.

[11] J. Bird, S. A. and Cohen-Cole, "The three-function model of the medical interview. An educational device," Advances in Psychosomatic Medicine., vol. 20, pp. 65–88, 1990.

[12] H. de Haes and J. Bensing, "Endpoints in medical communication research, proposing a framework of functions and outcomes," Patient Educ. Couns., vol. 74, pp. 287–294, March 2009.

[13] Z. Ben-Sira, "Affective and instrumental components in the physician-patient relationship: An additional dimension of interaction theory," J. Health Soc. Behav., vol. 21, pp. 170–180, June 1980.

[14] M. K. Buller and D. B. Buller, "Physicians' communication style and patient satisfaction," J. Health Soc. Behav., vol. 28, pp. 375–388, December 1987.

[15] B. M. Korsch and V. F. Negrete, "Doctor-patient communication," In Physician-patient Communication: Readings and Recommendations, G. Henderson, Ed. Springfield, IL: Charles C. Thomas, pp. 29–40, 1981.

[16] S. J. Williams and M. Calnan, "Convergence and divergence: Assessing criteria of consumer satisfaction across general practice, dental and hospital care settings," Soc. Sci. Med., vol. 33, pp. 707–716, 1991.

[17] R. C. Wasserman, T. S. Inui, R. D. Barriatua, W. B. Carter, and P. Lippincott, "Pediatric clinicians' support for parents makes a difference: An outcome-based analysis of clinician-parent interaction," Pediatrics, vol. 74, pp. 1047–1053, December 1984.

[18] H. S. Friedman, M. R. DiMatteo, and A. Taranta, "A study of the relationship between individual differences in nonverbal expressiveness and factors of personality and social interaction," J. Res. Pers., vol. 14, pp. 351–364, September 1980.

[19] J. Bensing, "Doctor-patient communication and the quality of care," Soc. Sci. Med., vol. 32, pp. 1301–1310, 1991.

[20] A. Mehrabian, "Communication without words," Psychol. Today, vol. 2, pp. 52–55, 1968.

[21] R. L. Birdwhistell, Kinesics and Context: Essays on Body Motion Communication. Philadelphia: University of Pennsylvania Press, 1970.

[22] H. S. Friedman "Nonverbal communication between patients and medical practitioners," J. Soc. Issues, vol. 35, pp. 82–99, Winter 1979.

[23] M. R. DiMatteo, L. M. Prince, and R. Hays, "Nonverbal communication in the medical context: The physician-patient relationship," In Nonverbal Communication in the Clinical Context, P. D. Blanck, R. Buck, and R. Rosenthal, Eds. University Park, PA: Pennsylvania State University Press, pp. 74–98, 1986.

[24] Microsoft Emotion API. [Online]. Available from: https://azure microsoft.com/en-us/services/cognitive-services/emotion/ 2018.07.06

Implantation of Elongated Silicon Neural Probe Array in Rat Cortex

Mohamad Hajj-Hassan, Rayan Fayad, Soumaya Berro
Houssein Hajj-Hassan
Department of Biomedical Engineering
International University of Beirut
Beirut, Lebanon
Email: mohamad.hajjhassan@b-iu.edu.lb,
rayanfayyad@gmail.com, soumaya.berro@b-iu.edu.lb,
houssein.hajjhassan@b-iu.edu.lb
Sam Musallam
Department of Electrical and Computer Engineering
McGill University
Montreal, Canada
Email: sam.musallam@mcgill.ca

Vamsy P. Chodavarapu
Department of Electrical and Computer Engineering
University of Dayton
Ohio, USA
Email: vchodavarapu1@udayton.edu

Hassan M. Khachfe
Lebanese Institute for Biomedical Research and
Application (LIBRA)
Lebanese International University
Beirut, Lebanon
Email: hassan.khachfe@liu.edu.lb

***Abstract*— Neural microprobes represent an important component of neural prosthetic systems where implanted microprobes record the electro-potentials generated by specific thoughts in the brain and convey the signals to algorithms trained to interpret these thoughts. Here, we present novel elongated multi-site neural probe that can reach depths greater than 10mm. We hypothesize that reaching such depth allows the recording of cognitive signals required to drive cognitive prosthetics. The impedance of the recording sites on the probes was in the order of 500 kΩ at 1 kHz, which is consistent with probes used for neurophysiological recordings. We implanted the elongated probe in rats and showed that the elongated probes are capable of recording spikes from various recording sites.**

Keywords-Cognitive neural prosthetics; Brain machine interfaces; Porous silicon; Microprobes.

I. INTRODUCTION

Brain Machine Interfaces (BMIs) have the potential to improve the lives of paralyzed patients by allowing them to use their neural activity to operate computers, robots, or even their own limbs [1] [2]. BMIs are designed to function in real time and benefit from real or simulated feedback. The development of BMIs as a direct communication pathway between the brain and external devices has generated novel methods and techniques to interface with and to study the brain. A BMI platform is comprised of 1) a system to record neural signals, 2) algorithms to interpret the neural signals and 3) the device to be controlled. In this paper, we focus on recording platforms composed of multiple probes implanted in the brain. These platforms must be biocompatible, must be designed to minimize the short term and long-term trauma inflicted during and after insertion. The probes must also be long enough to reach variable depths. Thus, probes must be made durable without increasing their width. Implantable probe arrays have traditionally been metal microprobes [3]-[5]. However, these have been recently supplanted by silicon probes [6]-[10]. Implanting probes into the brain elicits a tissue response that degrades the recorded signals. Regardless of substrate, probe design must curtail this response to ensure long-term recording.

Relative movement of the probes within the brain causes long-term tissue response due to the difference in mechanical properties between the probes and the neural tissue [11]-[13]. This process is exacerbated by arrays implanted deep in the brain due to their longer moment arm. Silicon probes can be made thin enough to increase compliance in the brain but without some rigidity, thin probes cannot penetrate neural tissue. Devices to assist implantation have been tested but may still cause neural damage during insertion and can only be used for surface arrays [14] [15].

We previously researched methods to develop implantable arrays made from silicon that can record signals from areas that are 6.5mm deep in the brain [16] [17]. Considerable progress in the design and fabrication of elongated silicon probes that can reach depths in the brain required for our applications were made. Silicon probes were reinforced probes with metallic structures making them more stable [10]. The rest of the paper is structured as follows: Section 2 details the fabrication process of the proposed probes. In section 3, the characterization and testing of the neural probes are presented. Section 4 is the conclusion where a summary of the work, main challenges, lessons learned, and future work are presented.

II. MICROFABRICATION PROCESS

The microfabrication process for the non-porous probe array has been described in detail in our previous publication [17]. In the current work, the microfabrication process follows similar steps at the beginning to form the probe array structure. During the end of the process, we perform additional processing to obtain porous probes. In brief, the microfabrication process begins with dicing a 50 µm thick 4" diameter double side polished silicon wafer (boron doped, resistivity of 20 ohm-cm and <100> oriented) into small square pieces by using a dicing saw.

Metal layers of titanium (adhesion layer, 500 nm thick) and gold (conducting layer, 750 nm thick) are then deposited by sputtering process on the silicon wafer, as depicted in Figure 1(a). The gold and titanium layers are photolithographically patterned and wet etched, with solutions of 1:2:10 I2:KI:H2O and 20:1:1 H2O:HF:H2O2, respectively, to define the recording pads, interconnects between the recording sites, and bonding pads as illustrated, in Figure 1(b). The silicon substrate was then patterned using photolithography and etched using isotropic xenon difluoride (XeF2) dry etching system to form the probe structures, as illustrated in Figure 1(c).

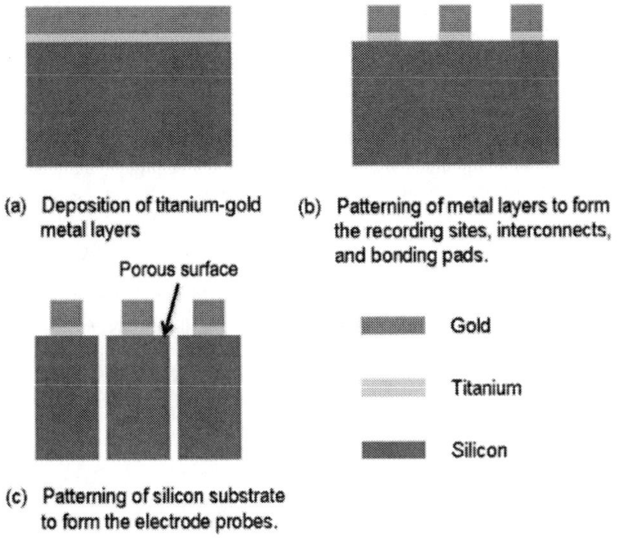

Figure 1. Schematic illustration of the steps involved in the fabrication process of the neural probe array (The Figure is not to scale).

The photoresist mask used in the previous step is removed and the silicon probe array is exposed etched using isotropic XeF2 dry etching system to form porous surfaces on the probe array. The formation of porous silicon using XeF2 dry etching system is described in our previous publications [18]-[20]. A 2μm thick layer of parylene-C, a biocompatible material widely used for coating a wide variety of implantable biomedical devices, such as pacemakers and silicon [21] and metal-wire neural probes [3], is conformably deposited, at room temperature using a Chemical Vapor Deposition (CVD) system, on the top side of the probe. It is mainly used to insulate the interconnects between the recording sites and bonding pads. Openings to expose the recording sites (used to measure the neural electrical activities) and the bonding pads (for wire-bonding to an external Printed Circuit Board (PCB) for read-out) were defined with photolithography. The exposed parylene-C is etched with oxygen Plasma Ashing System (PVA TePla Inc., Model: 200).

Figure 2(a) is a close-up view of the probe clearly showing the chisel-shaped tip, the titanium-gold recording site, and interconnects. Figure 2(b) is a magnified top view of the base of the probe showing the bonding pads used to connect the recording sites to external circuitry.

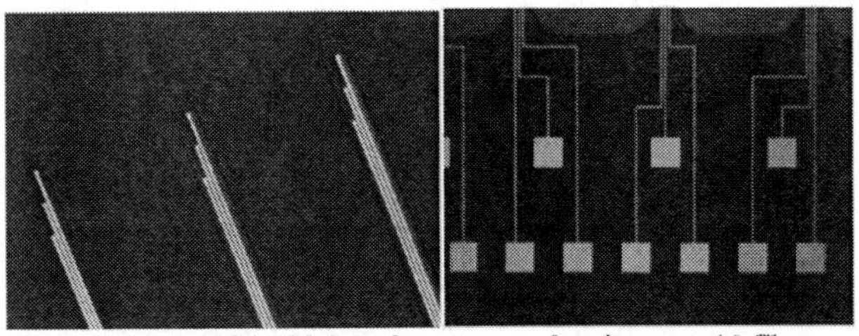
Figure 2. Images of the fabricated porous neural probe array. (a) Close-up image of the probe probes with recording sites; (b) Bonding pads.

The recording probe impedance is critical in the design of neural probes and is dominated by the size of the recording probe site.

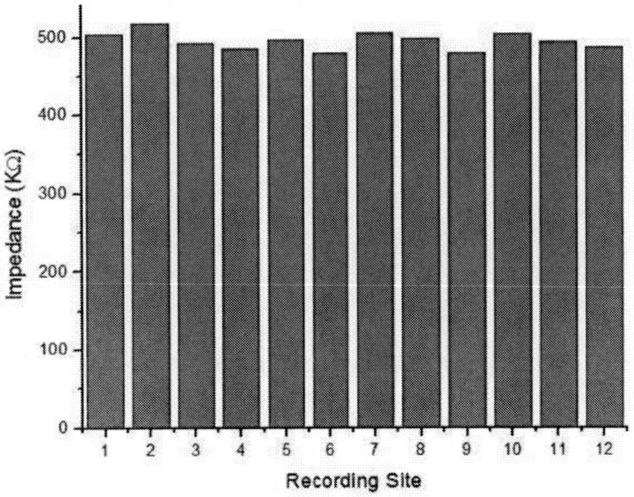

Figure 3. Impedance of the recording sites at 1 KHz.

The values of the impedance of the recording sites are obtained by recording the electric current while submerging the neural probes in saline solution and injecting current through them. Figure 3 shows the measured impedance for the twelve recording sites forming the current neural probe at 1 kHz and was found to be approximately 500 kΩ, a suitable impedance to record both single neuron activity and Local Field Potentials (LFPs) [3].

III. Testing and Results

The developed neural probe was tested in the barrel cortex of a rat. All procedures were approved by the McGill University Animal Care Committee on May 18, 2010 and were also in compliance with the guidelines of the Canadian Council on Animal Care. The protocol number is 5314. A Sprague–Dawley rat was handled for several days before the surgery in order to accustom it to handling by the investigators. We tested the probe array in the barrel cortex of a rat using a procedure described previously in [3]. Our target was subcortical nuclei. The probe was centered at 3 mm lateral and 2 mm anterior of the bregma and lowered 100-200-micron steps to a depth of 5 mm. The dura of the rat was not dissected prior to silicon probe insertion. A thin silver wire was placed under the skin and attached to a screw in the skull for use as an additional ground.

Spiking activity was recorded using a multi-channel acquisition processor (MAP, Plexon Inc., Dallas, TX, USA) where single units were isolated online using time–voltage windows and their timing and spike waveforms stored on computer.

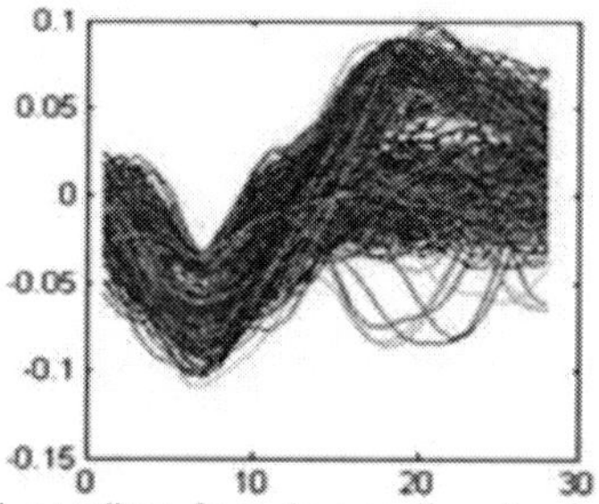

Figure 4. Neural recordings from six recording sites from the developed neural probe array.

Figure 4 shows recordings from one recording site of the developed neural probe array. Action potential waveforms from multiple neurons are visible in the signal. Multiple implantations are planned to gauge longevity of recordings. The rat was given pain, anti-inflammatory and antibiotic medication, as directed by our protocol and the McGill veterinarian. The rat was allowed to recover for at least 1 week before recordings were performed.

IV. Conclusion

The goal of the work reported here is to prove the functionality of the developed neural probe in accurately recording neural activity. We described a novel methodology to fabricate elongated multi-site neural microprobes. The probes were inserted through the pia and cortex of live rats to test their penetration ability. Results show a full insertion of the probes was successful without any bending, buckling, or breakage. Spiking activity was recorded simultaneously from the probes. The greatest challenge facing the use of neural probes is the resultant tissue response to the neural injury which effects the long-term functioning of the implanted neural probe. Future work will focus on improving the biocompatibility of these probes.

Acknowledgment

We are grateful for the financial support given by Natural Sciences and Engineering Research Council (NSERC) of Canada, Canadian Institutes for Health Research (CIHR), and the Canada Research Chair program. We would like to acknowledge the assistance of the McGill's Nanotools and Microfabrication Laboratory in preparing the described neural microprobe samples.

References

[1] R. A. Andersen, E. J. Hwang, and G. H. Mulliken, "Cognitive Neural Prosthetics," *Annual Review of Psychology,* vol. 61, pp. 169-190, 2010.

[2] J. Bartels, *et al.*, "Neurotrophic electrode: Method of assembly and implantation into human motor speech cortex," *Journal of Neuroscience Methods,* vol. 174, pp. 168-176, 9/30/ 2008.

[3] A. W. Bridges, *et al.*, "Chronic inflammatory responses to microgel-based implant coatings," *J Biomed Mater Res A,* vol. 94, pp. 252-8, Jul 2010.

[4] D. P. O. Brien, T. R. Nichols, and M. G. Allen, "Flexible microelectrode arrays with integrated insertion devices," in *Micro Electro Mechanical Systems, 2001. MEMS 2001. The 14th IEEE International Conference on*, 2001, pp. 216-219.

[5] P. K. Campbell, K. E. Jones, R. J. Huber, K. W. Horch, and R. A. Normann, "A silicon-based, three-dimensional neural interface: manufacturing processes for an intracortical electrode array," *IEEE Trans Biomed Eng,* vol. 38, pp. 758-68, 1991.

[6] M. C. K. Cheung, P. J. R. Roche, M. Hajj-Hassan, A. G. Kirk, Z. T. Mi, and V. P. Chodavarapu, "Controlling optical properties and surface morphology of dry etched porous silicon," *Journal of Nanophotonics,* vol. 5, May 13, 2011.

[7] K. L. Drake, K. D. Wise, J. Farraye, D. J. Anderson, and S. L. Bement, "Performance of Planar Multisite Microprobes in Recording Extracellular Single-Unit Intracortical Activity," *IEEE Transactions on Biomedical Engineering,* vol. 35, pp. 719-732, Sep 1988.

[8] M. Hajj-Hassan, M. Cheung, and V. Chodavarapu, "Dry etch fabrication of porous silicon using xenon difluoride," *Micro & Nano Letters,* vol. 5, pp. 63-69, Apr 2010.

[9] M. Hajj-Hassan, M. C. Cheung, and V. P. Chodavarapu, "Ultra-thin porous silicon membranes fabricated using dry etching," *Micro & Nano Letters,* vol. 6, pp. 226-228, Apr 2011.

[10] M. Hajj-Hassan, V. Chodavarapu, and S. Musallam, "NeuroMEMS: Neural Probe Microtechnologies," *Sensors,* vol. 8, pp. 6704-6726, Oct 2008.

[11] M. Hajj-Hassan, V. Chodavarapu, and S. Musallam, "Reinforced silicon neural microelectrode array fabricated using a commercial MEMS process," *Journal of Micro-Nanolithography Mems and Moems,* vol. 8, p. 033011, Jul-Sep 2009.

[12] M. Hajj-Hassan, V. P. Chodavarapu, and S. Musallam, "Microfabrication of ultra-long reinforced silicon neural electrodes," *Micro & Nano Letters,* vol. 4, pp. 53-58, Mar 2009.

[13] M. Hajj-Hassan, M. Khayyat-Kholghi, H. F. Wang, V. Chodavarapu, and J. E. Henderson, "Response of murine bone marrow-derived mesenchymal stromal cells to dry-etched porous silicon scaffolds," *Journal of Biomedical Materials Research Part A,* vol. 99a, pp. 269-274, Nov 2011.

[14] N. G. Hatsopoulos and J. P. Donoghue, "The science of neural interface systems," *Annu Rev Neurosci,* vol. 32, pp. 249-66, 2009.

[15] W. He, G. C. McConnell, T. M. Schneider, and R. V. Bellamkonda, "A novel anti-inflammatory surface for neural electrodes," *Advanced Materials,* vol. 19, pp. 3529-3533, Nov 5 2007.

[16] J. F. Hetke, D. J. Anderson, and K. D. Wise, "Design ranges for silicon multichannel neural probes," *Proceedings of the*

18th Annual International Conference of the IEEE Engineering in Medicine and Biology Society, Vol 18, Pts 1-5, vol. 18, pp. 266-267, 1997.

[17] B. Howell and W. M. Grill, "4 - Design of electrodes for stimulation and recording A2 - Kilgore, Kevin," in *Implantable Neuroprostheses for Restoring Function*, ed: Woodhead Publishing, 2015, pp. 59-93.

[18] C. H. Thompson, M. J. Zoratti, N. B. Langhals, and E. K. Purcell, "Regenerative Electrode Interfaces for Neural Prostheses," *Tissue Engineering Part B-Reviews,* vol. 22, pp. 125-135, Apr 1 2016.

[19] J. N. Turner, W. Shain, D. H. Szarowski, M. Andersen, S. Martins, M. Isaacson, *et al.*, "Cerebral astrocyte response to micromachined silicon implants," *Exp Neurol,* vol. 156, pp. 33-49, Mar 1999.

[20] R. Wadhwa, C. F. Lagenaur, and X. T. Cui, "Electrochemically controlled release of dexamethasone from conducting polymer polypyrrole coated electrode," *J Control Release,* vol. 110, pp. 531-41, Feb 21 2006.

[21] J. C. Williams, R. L. Rennaker, and D. R. Kipke, "Stability of chronic multichannel neural recordings: Implications for a long-term neural interface," *Neurocomputing,* vol. 26-7, pp. 1069-1076, Jun 1999.

Role of School Administration in Controlling Communicable Diseases

The Case of Influenza in Some West Bekaa (Lebanon) Area Schools

Walaa H. Sweid
School of Education
Lebanese International University (LIU)
Beirut, Lebanon
Email: 21230291@students.liu.edu.lb

Ali Al Khatib, Houssein Hajj-Hassan, Mohamad Rached
School of Arts and Sciences
International University of Beirut (BIU)
Beirut, Lebanon
Email: {ali.alkhatib; houssein.hajjhassan; mohamad.rached}@b-iu.edu.lb

Hassan M. Khachfe
Business, Educational, & Medical Optimization Research Institute (BE-MORE)
Lebanese International University (LIU)
Nabatieh, Lebanon
Email: hassan.khachfe@liu.edu.lb

***Abstract*–Academic performance and students' success are closely related to school attendance based on the evidence provided by literature. Influenza, an infectious disease, is one of the factors that prevent the students from attending classes. To study the role of school administration in controlling and preventing this disease and its effect on the students' performance, a field study was conducted in the schools of West Bekaa, Lebanon. Data was collected from private and public schools, in addition to the data collected from the Lebanese Ministry of Public Health. The findings revealed that influenza is one of the important reasons behind students' absence, which affects their performance. Immunization and general infection-control measures can be organized by the schools' administration to control and prevent the spread of this disease.**

Keywords-influenza; communicable diseases; controlling measures; immunization; absenteeism; administration.

I. INTRODUCTION

Influenza is a pandemic disease that spreads worldwide. The Centers for Disease Control and Prevention (CDC) and global health partners record that between 291,000 and 646,000 people are dying each year from seasonal influenza [1]. Unfortunately, schools in the region of Bekaa, Lebanon, are not aware of the risks of influenza and they consider it as a regular infection. Schools' administration can play an important role in controlling and preventing the spread of this disease among students and staff and, therefore, their family members.

High class attendance is correlated with better school performance. Absence from school is one of many factors that affect students' academic success [2]. School absence could be a common source of knowledge to monitor and detect the early propagation of diseases in schools especially in the case of influenza [3]. Students in schools are exposed to many communicable diseases that spread easily among them due to direct contact, so an absenteeism reporting system is needed in order to control the prevalence of the disease and control students' absenteeism [3].

Kids of school age are the ones who transmit influenza the most in society [4]. As stated by American Academy of Pediatrics [5], the flu is a communicable disease caused by an influenza virus which is extremely contagious and causes serious symptoms, which may end in hospitalization or death. In addition to affecting the respiratory system, it affects the whole body.

The remainder of this paper is organized as follows: Section II reviews the literature. Section III describes the methodology used. Section IV presents the results. Section V discusses the obtained results. Section VI sets the conclusion. Acknowledgement and references close the article.

II. LITERATURE REVIEW

Schools have a major role in controlling and preventing the transfer of Communicable Diseases (CDs) among students [6]. According to Michigan Department of Education [7], these diseases can transfer through fecal, oral and respiratory contact, direct skin-to-skin contact, indirect contact with contaminated objects and surfaces and blood borne contact. They also listed the most common communicable diseases that can affect students in school such as: influenza, common cold, diarrheal illness, *E-coli*, hand- foot and mouth disease, head lice, pink eye, ring worm, chickenpox, tuberculosis, measles and many others.

CD, such as influenza, affects both the global health and the economy if they are not controlled as fast as possible [8]. So non-pharmaceutical intervention strategies, such as: "wearing masks, continuous hand washing and school closure" should be done in order to control the spread of the diseases [8]. School closure is one of these intervention strategies because children can be easily affected and schools are the most places where children get contact with each other. After applying the strategy of school closure, the results showed that closing schools for several weeks can decrease the spread of disease but in a very low percentage because the patients are still active in the community outside the school. Also, this strategy leads to a negative economic impact, that's why other strategies should be applied inside the schools such as reducing the contact among students as much as possible [8].

Influenza is one of the most common communicable diseases that affect the students and the staff members [9]. It is a contagious respiratory illness caused by influenza viruses that infect the nose, throat and the lungs [10]. It can be transferred though coughing and sneezing when the distance between the infected and sick persons is less than one meter. The most effective way to prevent the infection could be to get immunized against the viruses through vaccination [10]. Antivirals, nutritional supplements such as vitamin D and physical barriers can also be used to decrease the viral infection among children [11].

Plaspohl *et al.* [12] stated that literature provides proof that school attendance correlates with educational performance and student success. Respiratory illness could be a conducive factor in high absences percentages of school students. The primary protection for respiratory illness is immunization. School-Located Influenza Vaccine (SLIV) programs offer larger access for college students to be vaccinated. A retrospective review of pre-existing information from four educational years was conducted to look at the link between SLIV participation and absence among students at eight public elementary faculties in Effingham County, Georgia. For the two SLIV years, results showed variations in average frequency of absences between SLIV and non-SLIV years also as between SLIV participants and nonparticipants. Implications for apply, embrace the potential for exaggerated herd immunity among students, which can conjointly reach different parties inside the college community and reception, so promoting overall healthy and successful students [12].

In this study, the aims were to analyze the role of school administration in controlling influenza in some Lebanese schools in West Bekaa region. Also the study aimed to assess the current situation of schools in coping with guidelines in terms of decisions, infection pathways, information sharing, communication, surveillance and to which extend they cooperate in this regard.

III. METHODOLOGY

A. *Research and instrument design*

As shown in the Figure 1, the research design used in this study is a mixed method that is called concurrent triangulation method (also called an integrative or convergent design) [13].

As per the conventions of mixed method notation, QUAN refers to quantitative methods and QUAL refers to qualitative methods. By using concurrent triangular design, described in the Figure 2, the researcher collects both qualitative and qualitative data, analyzes them and then interprets the results together for a better understanding of the topic [15].

In this research, a questionnaire was used – after being validated by the University's Institutional Research Board – to study the role of principals in controlling and preventing the prevalence of the influenza virus in West Bakaa's schools and its effect on students' performance. The questionnaire

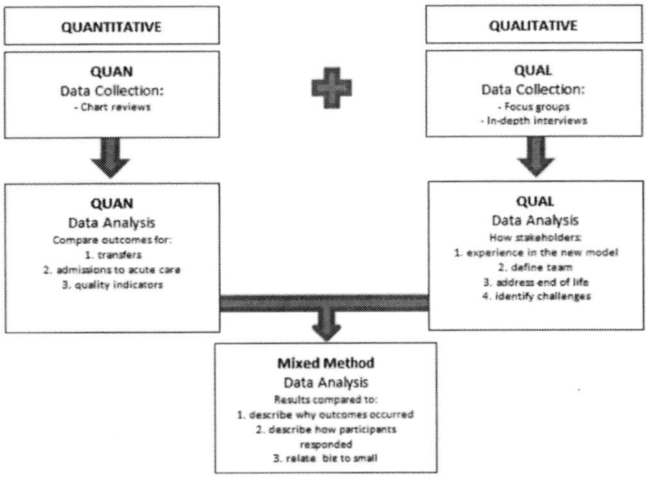

Figure 1. Concurrent Triangulation Design (adapted from [14])

questions were translated into Arabic to make sure that all the participants (principals) understand all the questions and answer accurately (validation).

The second source of data used in this study was the Ministry of Public Health (MoPH) database. All the information concerning the prevalence of the influenza virus in school aged children and its distribution according to the governorate was collected from the ministry's website [16].

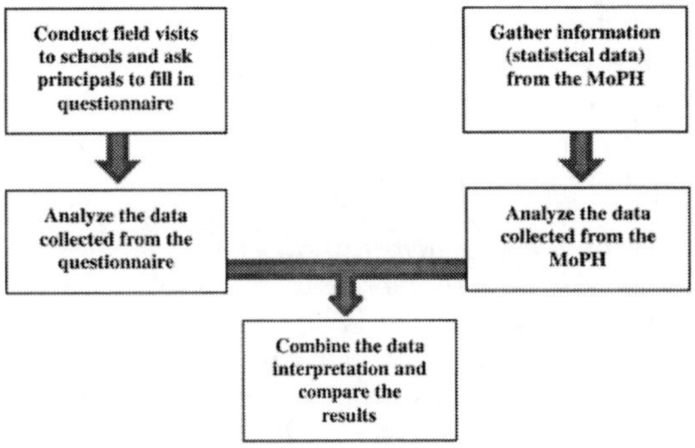

Figure 2. The design of the study used in this research (Concurrent Triangulation Design)

B. *Data collection*

Based on Figure 2, a questionnaire was distributed to 25 schools' principals in West Bekaa schools.

After that, a statistical data was collected from the Ministry of Public Health about the % of the infected people, their distribution according to the governorate and their age group. Data collected from the Ministry of Public Health covered the year 2017.

Then, the Concurrent Triangulation Design was applied in this study by analyzing all collected data and interpreting them. Finally, we combined and compared obtained results.

IV. Results

14.2% of the private schools' principals attended a workshop about communicable diseases under the title of H1N1 and health care in schools. 57.1% didn't attend any workshop, but they are interested in attending one, 14.2% didn't attend any workshop and they are not interested in that and 14.2 didn't answer this question.

Concerning the public schools, no one attended any workshop about this topic, but 81.8% are interested in attending such workshops and 18.1% didn't attend and they are not interested in that.

14% of the private schools' principals knew the number of the immunized students in the school (very low numbers) and the rest didn't know. 28.5% of them knew the number of the immunized staff (very low percentage) and the rest didn't. 63% of the public schools' principals knew the number of the immunized students and staff in the school, while the rest didn't.

85.5% of the private schools' principals stated the number of the students that absent due to influenza (the numbers varied between 1 to 10 students per day), 7% stated that no one absent and the rest didn't know or they didn't answer (Figure 3). 42% of them stated that the number of the staff absent due to influenza varied between 1 to 2 members per day, 14.2% stated that none of the staff members were absent due to influenza and the rest didn't know or they didn't answer (Figure 4). Concerning the public schools, 81% knew the number of the students absent due to influenza (the numbers varied between 2 to 10 students per day), 9% of them said that no one of them absent and the rest didn't know (Figure 5). 63.6% of the public schools' principals announced that one of the staff members was absent due to influenza per day and the rest announced that no one was absent due to that (Figure 6).

78% of the private schools' principals and 54% of the public ones considered that influenza was a serious disease and the rest of both didn't consider it serious. 92.8% of the private schools' principals and 81.8% of the public ones agreed that all the students and the staff members should be immunized against influenza and the rest didn't.

None of the private schools provided vaccines for students against influenza although they had a nurse in each of the schools, and 18.8% of the public schools did that, yet they all didn't have a nurse in any of the schools. 42.8% of the private schools and 27.2% of the public schools educated the students and the staff members about the symptoms of this virus, while the rest of both didn't.

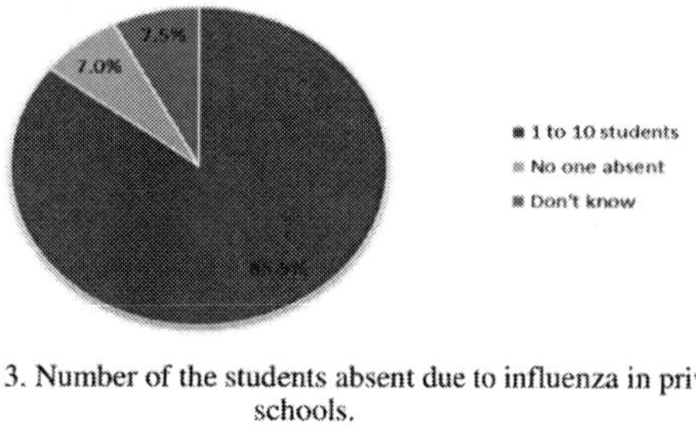

Figure 3. Number of the students absent due to influenza in private schools.

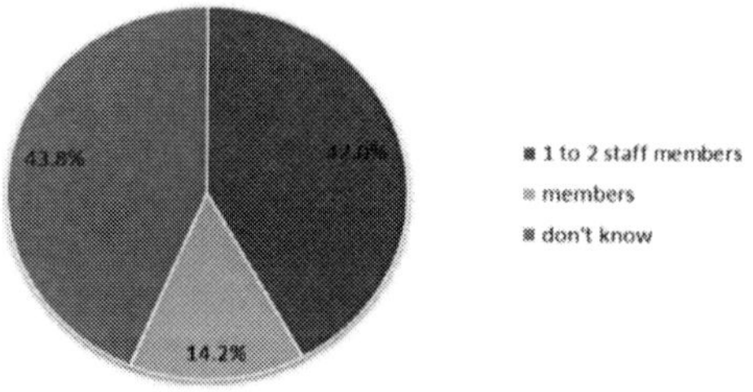

Figure 4. Number of the staff absent due to influenza in private schools.

All of the principals of private schools and 92.8% of the public ones were aware of the symptoms of the influenza and each one mentioned at least two of them. The answers varied between (fever, body aches, red eyes, runny nose, coughing, sneezing, sore throat and nasal congestion).

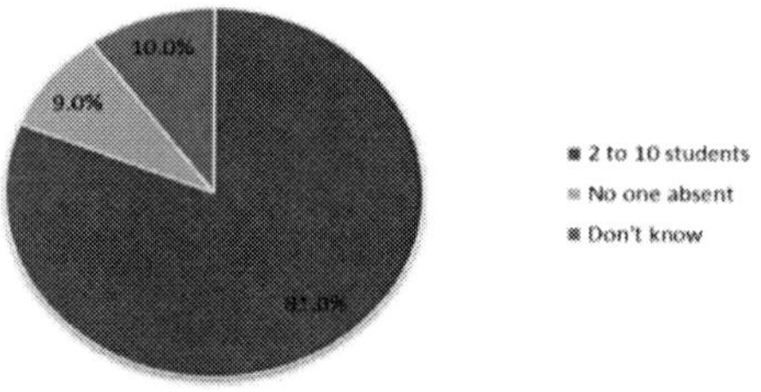

Figure 5. Number of the students absent due to influenza in public schools.

Principals were not aware enough about of the different types of influenza (A, B and C) and the one that affects students the most. 64.3% of the private schools' principals and 72.8% of the public ones were not aware about that. The rest answered that they were aware, but they incorrectly named the types such as swine and bird flu. Only few said that types A and B are the most common among students, they come from animals and have vaccine while C has no vaccine.

Figure 6. Number of the staff absent due to influenza in public schools.

All of the principals in both schools knew the ways of spreading of influenza. Each one of them mentioned at least one way. The answers were: touching, direct contact, sharing the same materials, coughing, sneezing, play grounds, toilets, low hygiene and drink tap water.

85.7% of the private schools' principals and 90.9% of the public schools' principals took decisions to control the spread of the influenza among the students in the school. Most of them focus on hygiene (personal hygiene, hand hygiene, the places where the students sit, and the materials that they use and share). The rest of the decisions were: eat healthy food, use tissues when coughing and sneezing and stay at home the first 48 hours when the symptoms appear. 7.1% of the private schools' principals said that they need to attend workshops concerning that. The rest didn't answer.

All the public schools and 92.8% of the private ones took measures concerning the infected students. Most of them called the parents and asked them to keep the student at home until full recovery. Around 4% of all exposed the patient to a specialist to receive the suitable medicine, advised the patient to drink more liquids, not to have direct contact with others and to get rest when feeling tired. Few of them asked the infected student to stay away and wear a mask. The decisions were taken by a health counselor and not by the administration in 9% of the public schools. 7.1% of the private schools' principals said that they need a workshop concerning that.

57.1% of the private schools' principals and 72.7% of the public ones believed that influenza can be prevented from spreading by following several ways such as: spread awareness, take the suitable decisions, remind them about the importance of the hygiene, eat healthy food, and provide vaccine. The rest of both didn't believe that influenza can be prevented from spreading.

57.1% of the private schools' principals and 72.7% of the public ones agreed that influenza affects the students' performance. Some said that it makes them loose their focus and concentration; the others said it is due to absence because students depend on working in the class and doing activities. Few of them said that it affects their performance if the teacher is absent for a long time due to influenza. The rest of both said that it has poorly or no effect.

To analyze the collected data, the results were compared to data retrieved from the website of the Ministry of Public Health. Table 1 shows the data collected from the Ministry of Public Health. Out of 1214 SARI cases from 11 sentinel sites in Lebanon 98% of the cases were tested (1186), where 9% of these cases were influenza (105). Out of these 9% cases 58% of them were influenza A (H3N2) (61). This % decreased to 42 % for influenza B (44). As for influenza A (H1N1), (H5N1), (H7N9) and others the % decreased to null (0%). As for the geographical distribution (Table 2), the % of influenza in Great North (North and Akkar) was the lowest (9%) followed by 26% in Great Bekaa, Beirut and Mount Lebanon (28%) and the highest recorded was in Great South (38%).

The % of influenza was 10% in age group between 5 and 14. This % increased slightly to 11% as the age group ranges between 50 and 64 and then it increased slightly to 13% as the age group ranges between 2 and 4. The influenza % increased to 16% as the age group ranges between 0 and 1. This % increased to 24% with a slight increase to 25% as the age group ranges between 15 and 49 and 65+ respectively.

As shown in Figure 7, at week 46, in 2016 and from week 10 to week 13, number of H3N2 cases was reported. However influenza B was reported but the number of cases was very low varying between 1% and 4% but with reported cases of H1N. However H3N2 appeared between week 49, 2016 and week 6, 2017 with a peak at week 52, 2016 with a value of 18% and type B also appeared between week 52, 2016 to week 8, 2017 with a peak at week 6, 2017 with a value of 10%.

TABLE I. THE PREVELANCE OF RESPIRATORY INFECTIONS IN SEASON 2016- 2017 [16]

Indicator	Nb	%
Sentinel	11	
SARI cases	1214	
Tested Cases	1186	98%
Influenza cases	105	9%
Influenza A	61	58%
Influenza A (H1N1)	0	0%
Influenza A (H3N2)	61	58%
Influenza A (H5N1)	0	0%
Influenza A (H7N9)	0	0%
Influenza A (other)	0	0%
Influenza B	44	42%

Between weeks 36 and 45 of year 2016, the number of SARI cases was very low, which varied between 0% and 25% however there was no influenza cases reported in this time interval. At week 45 of the same year, SARI cases increased slightly to become around 40% and influenza testing showed a slight increase to 2% and such value increased to 20% with the increase of SARI cases from 40% to 90% between weeks 50 and 52. However, such value decreased to nil with the decrease of SARI cases to 20% in the second week of 2017 while it peaked back to 28% at the 6th week of 2017 with the increase in SARI cases. But it decreased back with the decrease in SARI cases to become nil as long as number of SARI cases was low and below 20%.

Most principals in public schools and private ones were aware of severity and seriousness of the effect of influenza infection on students' performance and this was greatly obvious in the quantitative data since the percentage of the principals attended workshops on communicable diseases was very low (14.2% in private schools and null in public ones).

TABLE II. DISTRIBUTION OF PATIENTS BY GOVERNORATE (MOHAFAZA)

Mohafaza	Influenza %
Great North (North and Akkar)	9%
Great Bekaa (Bekaa & Baalbeck Hermel)	26%
Great South (South and Nabatieh)	38%
Beirut and Mount-Lebanon	28%

V. DISCUSSION

Regarding the question of being interested in attending workshops on communicable diseases and based on the measurements of principals who didn't attend any workshop but interested in, we can tell that both private and public schools show a high motivation and interest towards attending workshops that might come in handy in decreasing the severity of communicable diseases in schools. Thus, this can be shown by the high percentage that is scored in the data analysis (57.1 % for private schools and 81.8% for public ones).

86% of the principals of the private schools and 71.5% of the public ones didn't know the number of the immunized students against influenza in their schools. This is indeed a high percentage and consequently it shows that neither the principals of the private schools nor the public ones knew well the health status of their students. In addition, they aren't aware of the consequences that infection with influenza virus will reveal sooner on their achievement and academic performance.

A significant hope can be observed based on the analyzed data regarding the presence of the foundation of the prior knowledge concerning the main issue of our study. This can be shown in the percentage recorded by both private and public schools principals about the agreement on immunizing both the students and the staff members (92.8% and 81.8% respectively).

Despite the fact that the percentage of providing vaccine in public schools is low in comparison with the average percentage however, it is still acceptable compared with the private ones, which is null. Therefore, public schools took the initiative even if it is very slight to be observed as it is known regarding fighting against influenza virus but private ones didn't.

There is a low percentage in both private and public concerning raising the awareness among students and staff members about the symptoms of this virus (42.8% and 27.2% respectively).

Now, let us move towards discussing if the principals are aware of the symptoms of influenza. We could tell that they are totally aware of them. On the contrary, they aren't aware of the different types of influenza. This is absolutely referring back to the fact that they aren't enrolled in any workshops under this issue.

There is a high percentage which is observed in reference to both schools' principals with respect to taking decisions in controlling the spreading of influenza. This might be a good indicator if the taken decisions are right ones however, it might be not if vice versa.

Most of the principals of the schools followed the same strategies when dealing with infected students; however, few of them advocated different strategies that also came in handy in healing.

There is a great agreement by both principals on the fact that influenza can be prevented from spreading according to the percentage (57.1% for the private schools' principals and 72.7% for the public ones) whereas, the rest believe the opposite.

The majority of the principals of both schools agreed on the fact that influenza directly affects the performance and the achievement of the students whether the teacher is absent or the student is absent because of the influenza (57.1% for the private schools' principals and 72.7% for the public ones). Moreover, they also agreed that influenza affects their performance in different ways.

Based on the data collected from the Ministry of Public Health, influenza A and B are the most spread types of influenza and they are highly exposed to have Severe Acute Respiratory Infections.

Despite the fact that the percentage of influenza in Great Bekaa might appear low (26%) and with respect to the age

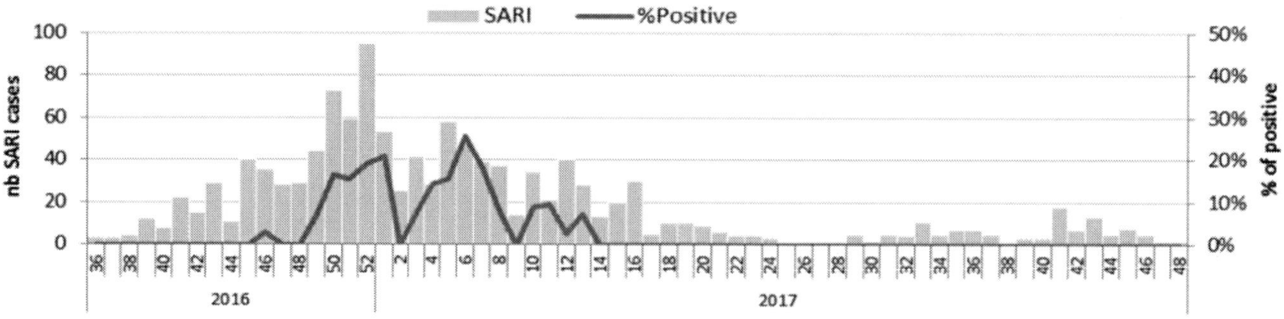

Figure 7. Distribution of Influenza Positive Patients by Week of Admission, 2016W36-2017W48 (Reproduced from [16])

group from 2-14 (23%) however, we can't deny its severity as one of the communicable diseases. Thus, it might be alerted percentage.

After analyzing Figure 3, " Distribution of Influenza positive patients by week of admission, 2016W36-2017W48" we concluded that influenza is mostly spread in winter season which is consistent with the results about influenza illness in England [17].

Our findings necessitate having a trained nurse in each school and establishing an awareness program to educate the school managers about the importance of management practices such as immunization programs, send sick students back to their homes, hand hygiene [18], school closure [19], and balanced meals with more fruits and vegetables.

VI. CONCLUSION

The results of the current study revealed that influenza is one of the important reasons behind students' absence. This affects their performance and achievements. Although most of the schools' principals were aware of some symptoms of influenza, they were not aware of the prevention measures to be taken to minimize transmission of the virus. Proper management of influenza in schools requires a plan to educate the schools principals about prevention measures and the ways to implement them.

ACKNOWLEDGMENT

The authors would like to thank all participating schools in the West Bekaa District and the Head of the Communicable Diseases Department at the MoPH, Dr. Atika Berry for their unconditional assistance.

REFERENCES

[1] M. Frellick, "Seasonal Flu Deaths More Common Worldwide Than Expected," Medscape Medical News, 2017. Available from: https://www.medscape.com/viewarticle/890082

[2] V. Kassarnig, A. Bjerre-Nielsen, E. Mones, S. Lehmann, and D. Lassen, "Class Attendance, Peer Similarity, and Academic Performance in a Large Field Study," PLOS ONE, 12(11): e0187078, Doi: 10.1731, 2017.

[3] S. Lawpoolsri, A. Khamsiriwatchara, W. Liulark, K. Taweeseneepitch, A. Sangvichean, W. Thongprarong, J. Kaewkungwal, and P. Singhasivanon, "Real-Time Monitoring of School Absenteeism to Enhance Disease Surveillance: A Pilot Study of a Mobile Electronic Reporting System, ' JMIR mHealth and uHealth, 2(2):e22, Doi:10.2196/mhealth.3114, 2014.

[4] P. Pannaraj, H. Wang, H. Rivas, H. Wiryawan, M. Smit, N. Green, G. Aldrovandi, J. Ridenhour, A. Braun, T. Teyrasse, and D. Goldsman, "Controlling the Spread of Disease in Schools," PLoS ONE, 6.(12): e29640, doi:10.1371, 2011.

[5] American Academy of Pediatrics, "Influenza Prevention and Control," 2015 [Online]. Available from: https://www.aap.org/en-us/Documents/disasters_dpac_InfluenzaHandout.pdf 2018.09.20

[6] Michigan Department of Education (MDE) and Michigan Department of Health and Human Services (MDH), Divisions of Communicable Disease & Immunization, "School Management of Measles," 2017.

[7] Michigan Department of Education (MDE) and Michigan Department of Health and Human Services (MDH), Divisions of Communicable Disease & Immunization, "Managing Communicable Diseases in Schools," 2016.

[8] B. J. Ridenhour, A. Braun, T. Teyrasse, and D. Goldsman, "Controlling the Spread of Disease in Schools," PLoS ONE, 2011. [Online]. Available from: https://doi.org/10.1371/journal.pone.0029640 2018.08.15

[9] M. De Perio, D. Wiegand, and S. Brueck, "Influenza Vaccination Coverage among School Employees: Assessing Knowledge, Attitudes, and Behaviors," J Sch Health, 84(9): 586–592, doi:10.1111/josh.12184, 2014.

[10] M. Moghadami, "A Narrative Review of Influenza: A Seasonal and Pandemic Disease," Iran J Med Sci, 42(1): 2–13, 2017.

[11] B. Lee, and M. Shah, "Prevention of Influenza in Healthy Children," Expert Rev Anti Infect Ther, 10(10): 1139–1152, doi:10.1586/eri.12.106, 2012.

[12] S. Plaspohl, B. Dixon, and A. Streater, "Impact of School Flu Vaccine Program on Student Absences," Journal of School Nursing, 1: 75-80, Doi:10.11711, 2014.

[13] V. Sousa, M. Driessnack, and I. Mendes, "An overview of research designs relevant to nursing: Part 1: Quantitative research designs," SciELO, 3: 1518-8345, Doi:10.1590, 2007.

[14] J. Creswell, and V. Plano Clark, "Designing and Conducting Mixed Methods Research," SAGE Publishing. ISBN: 1412975174, 9781412975179, 2011.

[15] F. Castro, J. Kellison, S. Boyd, and A. Kopak, "A Methodology for Conducting Integrative Mixed Methods Research and Data Analyses," National Institution of Health, 4(4): 342–360, doi:10.1177/1558689810382916, 2010.

[16] MOPH, "Lebanese Severe Acute Respiratory Infections (SARI) Sentinel Surveillance" 2017 [Online]. Available from: https://www.moph.gov.lb/userfiles/files/Esu_data/Esu_current year/sari_monthly.pdf

[17] R. Aldridge, A. Hayward, N. Field, C. Warren-Gash, C. Smith, R. Pebody, D. Fleming, and S. McCracken, "Are School Absences Correlated with Influenza Surveillance Data in England? Results from Decipher My Data—A Research Project Conducted through Scientific Engagement with Schools," PLoS ONE 11(3): e0146964, doi:10.1371/journal.pone.0146964, 2016.

[18] M. Talaat, S. Afifi, E. Dueger, N. El-Ashry, A. Marfin, A. Kandeel, and N. El-Sayed, "Effects of Hand Hygiene Campaigns on Incidence of Laboratory-confirmed Influenza and Absenteeism in Schoolchildren, Cairo, Egypt," Emerging Infectious Diseases, 17(4), 619–625, doi:10.3201/eid1704.101353, 2011

[19] N. Halder, J. Kelso, G. Milne, "Developing guidelines for school closure interventions to be used during a future influenza pandemic," BMC Infectious Diseases, 10:221, 2010. [Online]. Available from: http://www.biomedcentral.com/1471-2334/10/221 2018.09.08

Data Driven Medical Process Modelling for Privacy Protection in Care Pathways

Intidhar Essefi
University of Tunis el Manar, The Higher Institute of Medical Technologies of Tunis
Research Laboratory of Biophysics and Medical Technologies
Tunis, Tunisia
e-mail: essefi.intidhar@gmail.com

Hanen Boussi Rahmouni
University of Carthage,
Higher School of Communication of Tunis
Digital Security Research Unit
and
University of Tunis el Manar, The Higher Institute of Medical Technologies of Tunis
Tunis, Tunisia
e-mail: hanene.boussi@istmt.utm.tn

Mohamed Fethi Ladeb
Radiology Department
Kassab Orthopedics Institute
Manouba, Tunisia
e-mail: fethiladeb@hotmail.fr

Abstract—**In this article, we present a clinical pathway specification methodology for data driven medical process modelling. Our model takes into consideration patients privacy preservation. It gives special attention to shared medical documents structure and content. Furthermore, our model describes the different clinical tasks, typically included in hospitals care pathways. It also exposes the underlying shared patient data enclosed within the medical documents required by the clinical pathway subject to execution. This research aims first of all to identify, for each clinical task that requires data processing or sharing, the level of protection that data requires. For this, we suggest to extend existing business process modelling languages with special means highlighting medical data. As a second step, we aim to map each extracted data category to a set of privacy requirements as demanded by the Health Insurance and Accountability Act (HIPAA) legislation. This will ensure the respect of data protection requirements since a very early stage of Hospital Information Systems (HIS) design.**

Keywords-business process modelling; clinical pathways; data driven; HIPAA legislation; patient privacy; privacy requirements.

I. INTRODUCTION

In order to improve services' quality within a hospital environment, it is important to automate the underlining workflows of each clinical process adopted within the hospital. This needs the design and the implementation of business process models tailored for the concerned field. Medical healthcare is a multidisciplinary field. Its business processes and workflows are very complex. Throughout each medical process, several types of clinical information need to be circulated and treated within or without the hospital's boundaries. Medical data are produced, transmitted between medical departments and shared between healthcare professionals throughout the clinical pathways enforced by the hospital information systems in use. Several types of medical data documents are processed, including admission papers, insurance documents, prescriptions, confidential letters, medical images, imaging reports, biological reports, other types of medical reports, etc. All the mentioned clinical documents include diverse health information, among which we distinguish sensitive information that is considered as highly Protected Health Information (PHI). Personal healthcare information is not only used in healthcare practices and shared between healthcare professionals, but also in public practices and research activities such as public health surveillance and public health research. Public health practices and research present risks that are related to the unauthorized disclosure of PHI [1]. Therefore, it is crucial for healthcare organizations to ensure PHI protection and to preserve the privacy of individuals. Particularly, the individual's privacy protection is required by legislation, such as the Health Insurance Portability and Accountability Act (HIPAA) legislation [16][17] and the European Directive [18] on personal data protection. As a consequence, privacy requirements should be respected and ensured when designing systems and procedures for health data management.

Our approach is based on privacy by design, which means the implementation of privacy requirements since an early stage of healthcare information systems design with respect to carrying out clinical pathways. In this paper, we take the osteosarcoma clinical pathway as a case study to validate our approach. The details of our approach are as follows:

- Model medical care pathways as business processes that emphasise shared clinical data aiming to identify sensitive health information among them.
- Identify the privacy requirements and procedures for each type of sensitive health data identified within the business process representing each care pathway.
- Identify a clinical data model based on the business process modelling.
- Define the sensitive health information categories.
- Define the HIPAA legislation requirements to preserve the patient's privacy and confidentiality with regards to the use of their PHI.
- Model the clinical pathways based on business process modelling in order to extract the shared clinical documents between healthcare professionals.
- Identify the PHI underlining each process model.

- Define privacy requirements for PHI protection from any disclosure or misuse.

This paper is divided into sections as follows: in Section II, we present the related work; in Section III, we present the clinical pathway subject to study, as well as the clinical pathway modelling language of our choice. We adopt a data driven business process clinical pathway modelling approach. In Section IV, we present our clinical document architecture. In Sections V and VI, we define the privacy requirements for PHI, followed by results and discussion. In Section VII, we present the conclusion and future work.

II. Related work

In the literature, clinical pathways are textually and medically described. Their business processes and workflows are mostly detailed by doctors using textual description of the sequenced tasks. Due to the technological revolution in the medical field that includes medical information systems, several methods and business process modelling languages have emerged. This includes the Integration DEFinition language (IDEF) (V.0 and V.3), the Unified Modelling Language (UML) V.2.0 and the Business Process Model and Notation language (BPMN). Most of these technologies were also used to model clinical pathways [2].

BPMN is the most widely used and accepted language in medical process modelling thanks to its simple and high-level process construction. Clinical pathways processes are known as complex. This needs a transparency of the whole process elements such as structures, participants, tasks, roles, etc. Modelling care pathways in the form of clinical processes is considered as a solution to overcome its complexity and define its requirements with regards to patients and health care professionals. Therefore, medical processes models should be simple, transparent and understandable as much as possible [3][4].

Despite the importance of business process modelling in clinical pathways and efforts for processes' automation, few works are dealing with care pathways' automation. Most of them are relying on a business process-based modelling approach. Besides, there is some suggested BPMN extension implementation such as the Clinical Pathway (CP) extension of the BPMN called BPMN4CP which proposes an ontology based- extension for e-health process management. Other existing research works offer clinical textual description of the care pathways processes. In addition, other works were interested in analyzing systems behavior throughout business process-based modelling using UML, particularly, UML class diagram [5]-[7].

However, less effort was made in investigating approaches for clinical data modelling with special interest in privacy preservation. In this context, our work is addressed to the respect of privacy requirements since a very early stage of HIS design in order to ensure a protected rolling of data driven business processes that are clinical pathway-oriented.

III. Clinical pathways

Clinical pathways are acknowledged as complex processes due to the diversity of the participating entities (e.g., healthcare professionals and medical service providers). Throughout the literature exploration of the clinical pathways' modelling and automation, we extracted the main phases underlining care pathway processes. A generic clinical pathway begins by an admission phase in which the patient is allowed to get access to the care establishments. This is usually followed by a diagnosis phase: that describes the visiting of the consulting doctor and having clinical diagnosis performed. The treatment phase should then occur: after identifying the pathology in the second phase, the treating doctor identifies the treatment protocol. This clinical pathway ends with the follow-up phase which allows the involved practitioner to monitor and evaluate the effectiveness of the prescribed treatment or to control the pathology progression [8].

The clinical pathway is a set of processes and sub-processes in which one or more healthcare professionals participate. The business process modelling of clinical pathways allows to identify the tasks, the participants and their roles in the care pathway proceeding. Even the shared data between healthcare professionals may also be modelled and identified among a clinical pathway-oriented business process [9].

In the following sections, we will detail the clinical care pathway of osteosarcoma and describe a step by step methodology to model our clinical business process.

A. An Overview of Osteosarcoma Clinical Pathways

Osteosarcoma is a bone cancer. It most commonly reaches those aged from 10 to 30. A great part of this affected population concerns teenagers. Each year, from 800 to 900 people are estimated to be diagnosed with osteosarcoma in the United States. Osteosarcomas are primary malignant bone tumors. They can be classified according to cells' behavior under the microscope as high, intermediate or low grade [10].

Osteosarcoma clinical care pathways are characterized by their complex and multidisciplinary procedures with their difficult management facts. By Ferrante [3], the osteosarcoma first diagnosis starts with symptoms appearances like bone pain or soreness, a felt mass through the skin, swelling and redness, etc. During the clinical pathway diagnosis phase, while an osteosarcoma is suspected, some standard imaging exams must be performed. Once the osteosarcoma is confirmed and its malignancy is not excluded, a biopsy should be performed allowing the cancer staging. As a final checkup step, several imaging exams are performed to verify the existence of metastases. A percentage of 85% indicates that the most common metastases appear in the lung whereas the bone is considered as the second most common site of distant disease [11].

The osteosarcoma checkup and grading steps allow the choice of the treatment procedure which includes chemotherapy, radiation therapy and surgery operation. The identification of the right therapy protocol is based on biological analyses. To verify and evaluate the treatment efficiency, the patient has to be periodically followed-up. This osteosarcoma clinical pathway follow-up step is based on the performing of imaging exams in addition to biological analyses as needed [11]-[13].

The complexity of the osteosarcoma clinical pathway business process is due to the collaboration between healthcare professionals from several medical departments. This process can not be accomplished without clinical data sharing and transmission. For that, it is necessary to respect the applicable data protection regulation.

B. BPMN as Clinical Pathway Modelling Language

To model clinical pathways, we used the BPMN as a modelling language. It is the most widely used language in healthcare business process modelling. First, we explored the clinical healthcare pathways in the literature on description of clinical pathways. Then, we divided them into three main phases, diagnosis or check-up, treatment and follow-up, mentioned above. Throughout the studied clinical pathways, we present the osteosarcoma clinical pathway as a case study to illustrate our data driven clinical healthcare pathway model. In order to elaborate the clinical pathway data driven model, we used the common patterns and symbols of the BPMN modelling language [2].

C. Osteosarcoma Clinical Pathway Modelling Using BPMN

In order to identify the clinical data that may be transmitted and shared between healthcare professionals as required by standard care pathway specifications, we modelled the osteosarcoma clinical pathway in the form of a business process model. In this way, we could first identify the characteristics of performed clinical tasks. Then, we could highlight the data driven tasks for the chosen pathology. The sections below present the data driven clinical pathways of osteosarcoma for the check-up, the treatment and the follow-up phases respectively.

1) Osteosarcoma checkup clinical pathway

Fig. 1 presents the check-up phase of the osteosarcoma clinical pathway as well as the shared and transmitted clinical documents, ensuring the steps required by the care pathway definition. The osteosarcoma clinical pathway is complex and involves collaboration of healthcare providers and diverse medical services including radiology, biology, nuclear medicine, surgical units, etc., as shown in Fig. 1, Fig. 2 and Fig. 3. After the patient admission, a medical consultation takes place. According to the clinical examination, medical tests are performed to accomplish the diagnoses phase and precise the pathology Fig. 1.

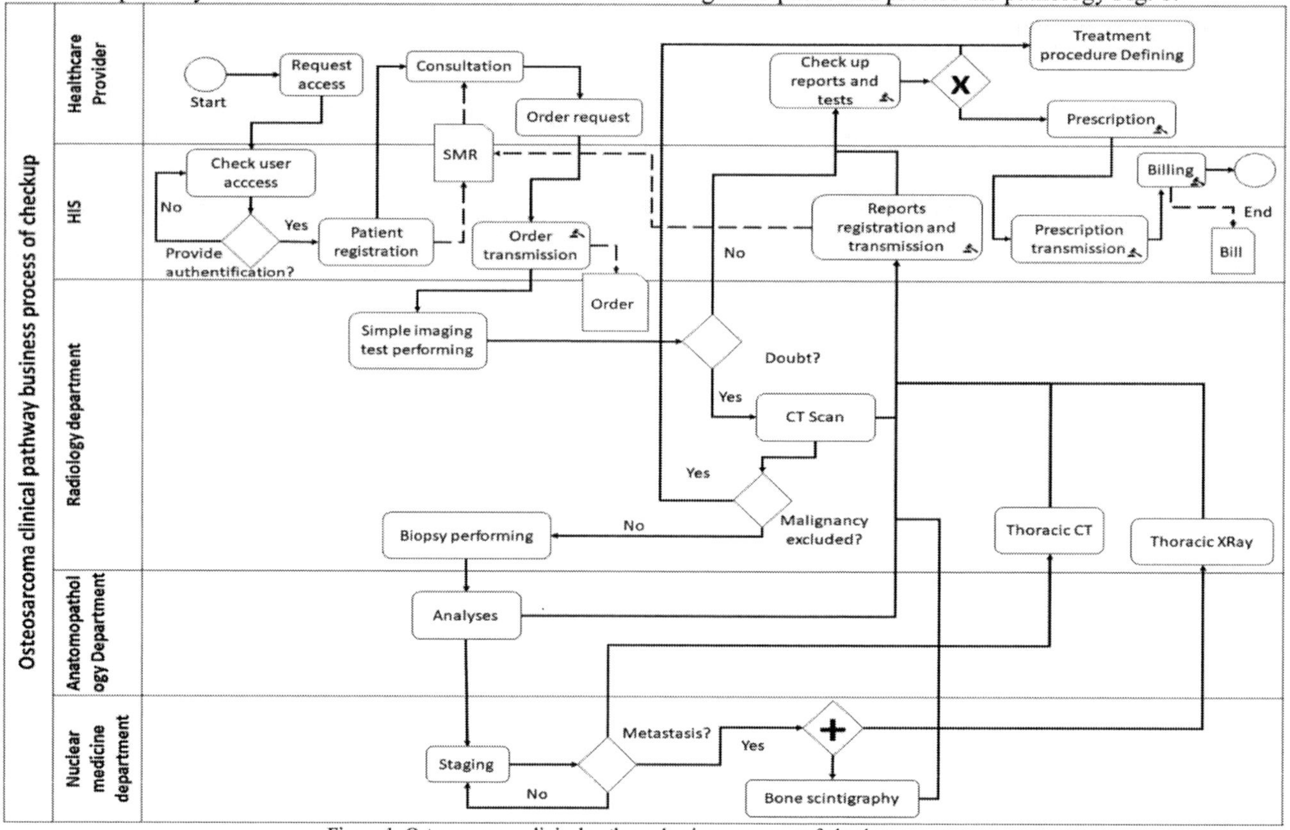

Figure 1. Osteosarcoma clinical pathway business process of checkup.

2) Osteosarcoma treatment clinical pathway

Based on tests' findings within the diagnosis phase, the doctor defines the treatment phase according to systems review by biological tests, audiogram hearing tests and heart tests. By Luetke [12], during the treatment, medical tests and clinical examination are performed to evaluate its effectiveness, as shown in Fig. 2.

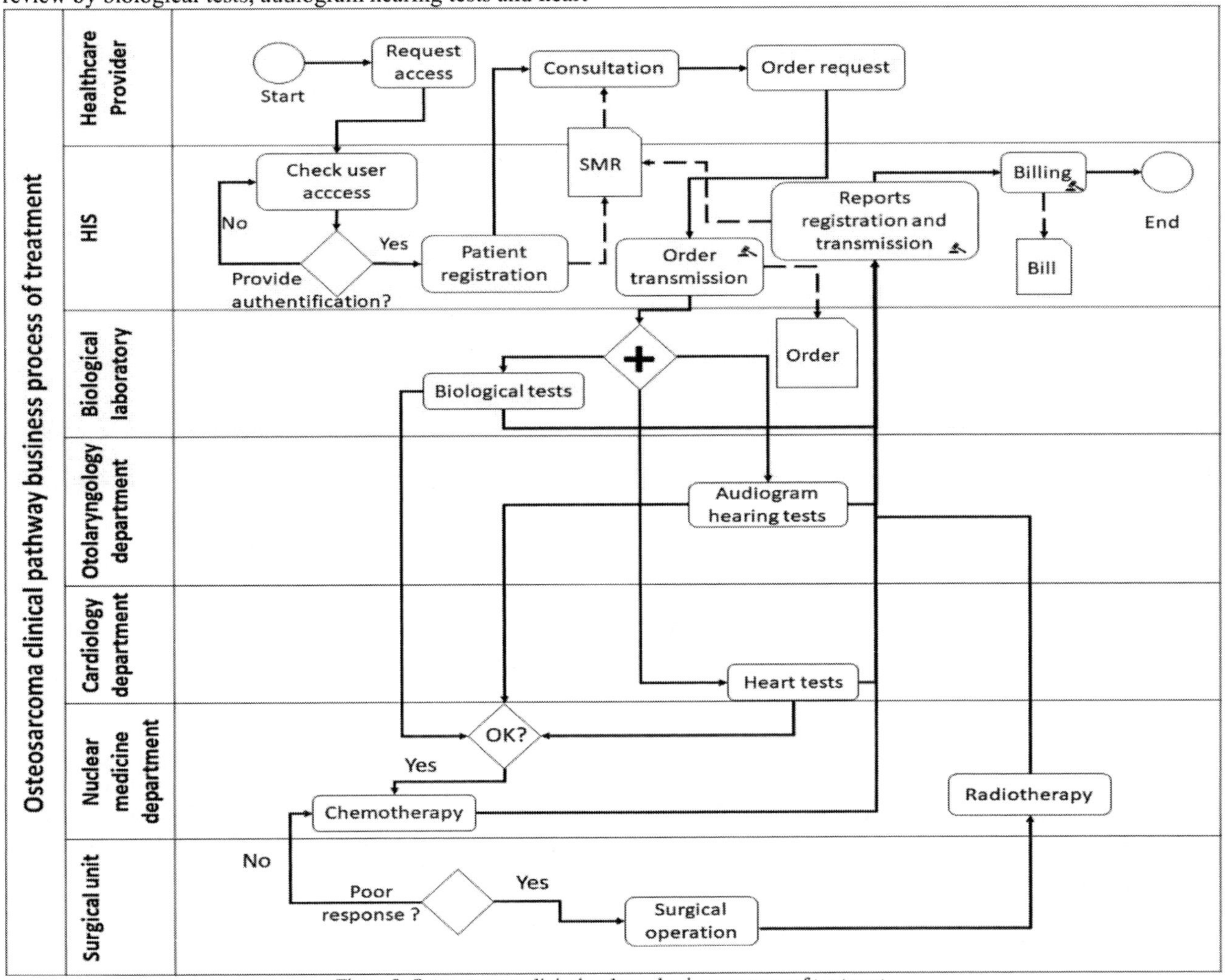

Figure 2. Osteosarcoma clinical pathway business process of treatment.

3) Osteosarcoma follow up clinical pathway

The last step is the following-up phase presented in Fig. 3. According to Paiolil [13], the doctor follows the patient health status by performing some tests and medical examination to check periodically the treatment effectiveness. Throughout the three phases of the osteosarcoma clinical pathway, diverse clinical documents are shared, transmitted and updated within the Shared Medical Record (SMR) ensuring the healthcare continuity.

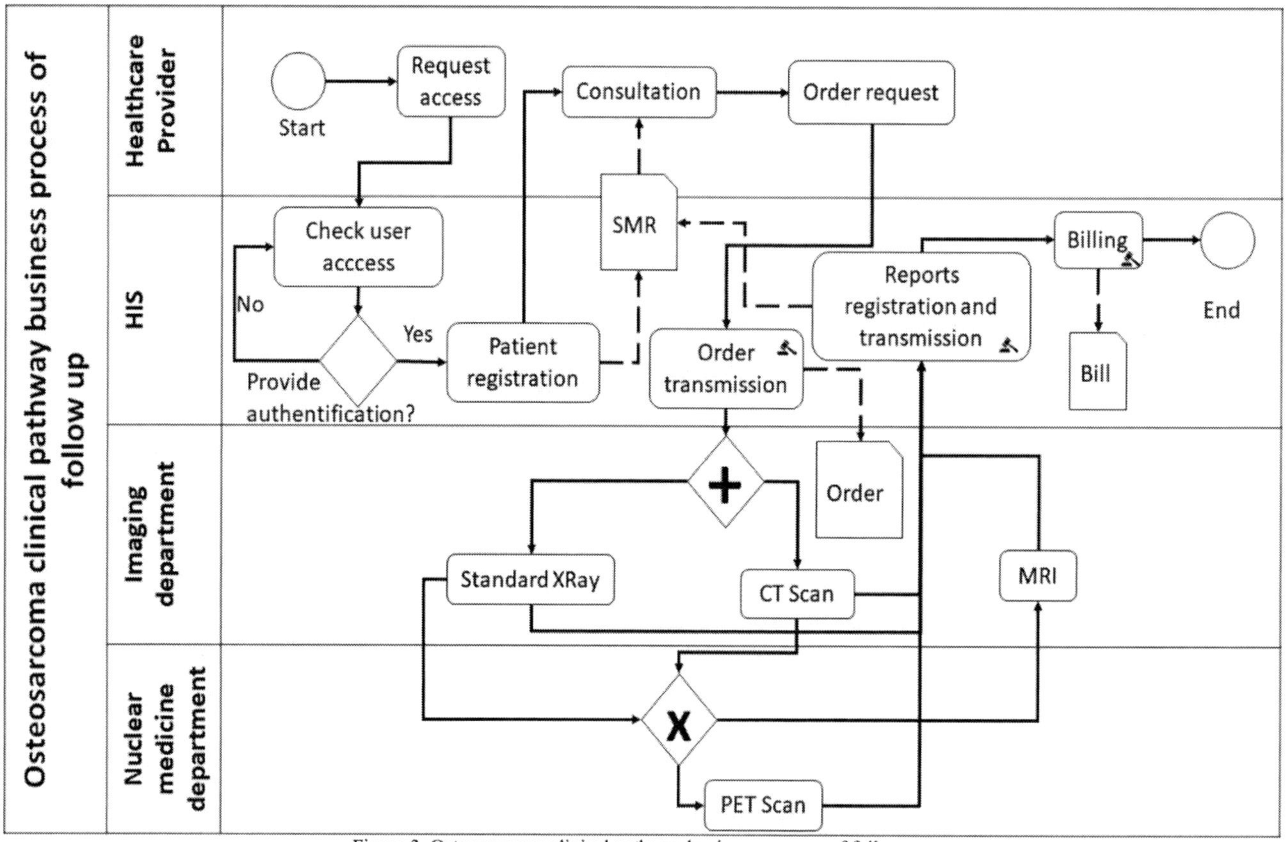

Figure 3. Osteosarcoma clinical pathway business process of follow up.

IV. SHARED CLINICAL DOCUMENT ARCHITECTURE

A patient's Electronic Health Record (EHR) must contain all types of clinical documents including their medical history record, discharge summaries, typical paper charts, mental status examinations and other medical reports, such as medical tests and operative reports.

Throughout a clinical business process, the EHR is transferred, updated and shared between healthcare professionals ensuring the continuity of care. Each clinical document included in the EHR contains medical data as it is required in the concerned healthcare establishment.

The general clinical document architecture is divided into documents, fragments and data. As shown in Fig. 4, clinical documents are composed of many fragments. They provide information about patients, procedures, practitioners, diagnosis, findings and appointments. The clinical shared documents' architecture model, illustrated in Fig. 4, could be adapted to another health care establishment, according to the used medical documents' structure in their boundaries. In each clinical document fragment, several medical data are found with specific properties which need the implementation of a privacy by design approach. This is dedicated to the PHI use and disclosure within the HIS. The identification and demographic fragments in clinical documents include PHI. Its use should obey to the data protection law principles and privacy requirements ensuring the PHI privacy and the security of the medical data in use [14] [15].

This could be applied by the use of security computerized methods (e.g. encryption, decryption, anonymization and pseudonymization) since an early stage of the design of the HIS.

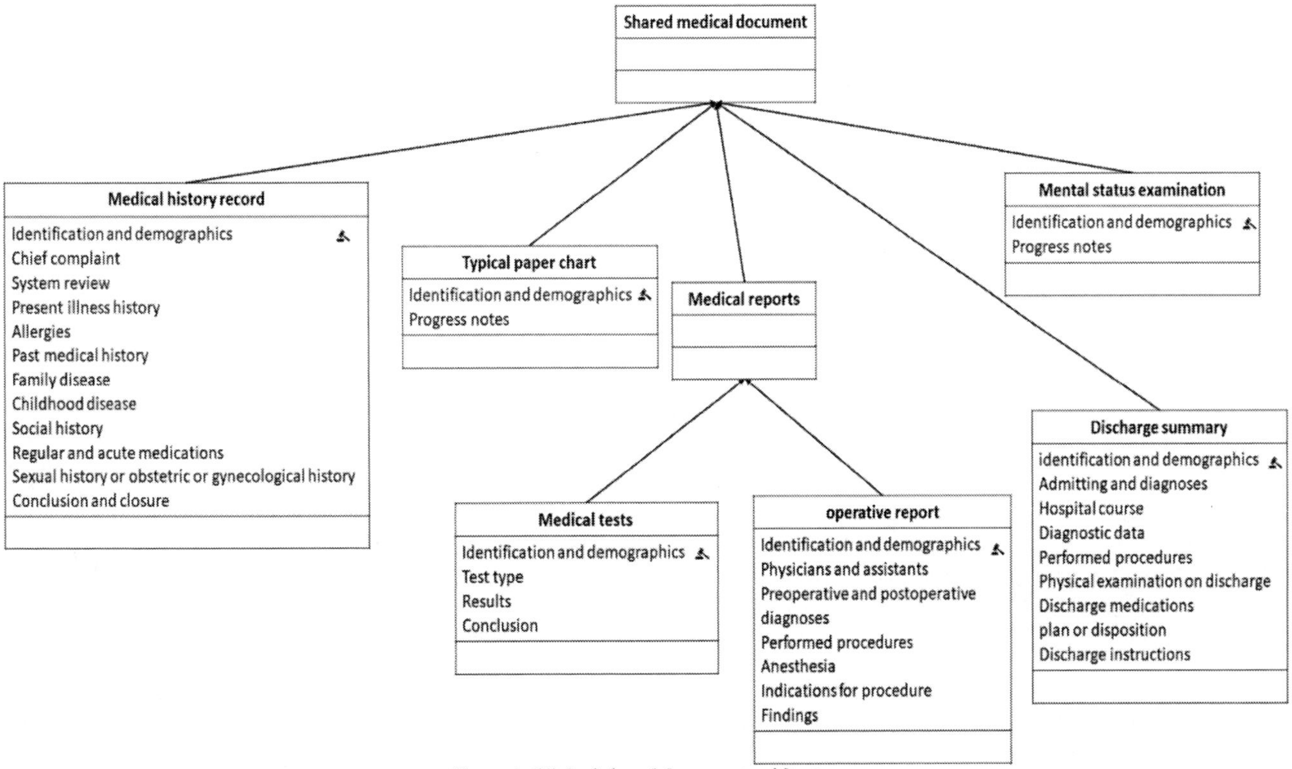

Figure 4. Clinical shared document architecture.

V. PRIVACY REQUIREMENTS FOR PHI PROTECTION

The use of clinical healthcare data is governed by many jurisdictions as it may present risks threatening a person's life and may affect both his privacy as well as his professional life. For this, Clinical data usage must be set for data protection principles. In particular, the following eight principles should be respected:

1- Lawfulness, fairness and transparency: personal data should be processed lawfully, fairly in a transparent way.
2- Purpose limitation: personal data should be processed for specific purposes.
3- Data minimization: personal data should be adequate, relevant and limited to the precise purposes.
4- Accuracy: personal data should be kept up to date.
5- Storage limitation: personal data should be kept for no longer than the necessary period for the purposes for which those data are processed.
6- Rights: people have the right to access their data and give permission for other entities to use or disclose them.
7- Integrity and confidentiality: personal data should be processed in a secure way. They should be protected also against any unauthorized or unlawful processing, accidental loss, destruction or damage.
8- International transfers: personal data should not be transferred outside countries [14].

International law frameworks, such as European directive and HIPPA for personal data protection are based on the previous data protection principles. The present work is developed with regard to Protected Health Information within HIPAA regulation. The HIPAA Privacy Rule is published by the department of Health and Human Services (HHS) to ensure health information privacy. The privacy rule is applied to covered entities as health plans, healthcare clearinghouses and the healthcare providers. It defines a set of rules in order to protect sensitive health information with respect to its use and disclosure. Sensitive health information is known as Protected Health Information (PHI). They are individually identifiable health information related to the patient's past, present and future physical or mental health conditions, the healthcare provision to the individuals and the past, present or future healthcare provision to individuals [1]. The individually identifiable health information includes demographic data and many common identifiers. PHI usage and disclosure are permitted without the patient's informed consent for some purposes and situations as to the individual, the treatment, payment and healthcare operations, opportunity to agree or object, incidence to an otherwise permitted use and disclosure or public interest and benefit activities as well as a limited data set for research, public health or healthcare operations purposes or when it is required by law. As for the not permitted PHI usage and disclosures, an individual's written authorization (consent) must be obtained [16].

In addition to permitted PHI use and disclosure, prohibited ones are defined in Privacy Rules. For example, genetic information is considered as PHI and they shall not be used or disclosed for underwriting purposes as well as the psychotherapy notes. Furthermore, PHI may not be sold by covered entities. The PHI use and disclosure must be limited to the minimum necessary. However, PHI may be used to create a non-individually identifiable health information or a de-identified information [16][17].

The HIPAA Privacy Rule also defines a set of PHI de-identification requirements in order to use and disclose it without the patient's authorization. A covered entity may de-identify PHI by removing the eighteen identifiers specified in the following list:

1. Names.
2. Addresses with all geographic subdivisions smaller than a State.
3. Dates except year (birthdate, admission and discharge date, date of death).
4. Telephone numbers.
5. Fax numbers.
6. Email addresses.
7. Social security numbers.
8. Medical record numbers.
9. Health plan beneficiary numbers.
10. Account numbers.
11. Certificate/license numbers.
12. Vehicle identifiers, serial numbers and license plate numbers.
13. Device identifiers and serial numbers.
14. URLs (Web Universal Resource Locators.
15. IP (Internet Protocol) address numbers.
16. Biometric identifiers (finger and voice prints).
17. Full face photographic images and any comparable images.
18. Any unique identifying number characteristic or code [16].

For the above identified PHI use and disclosure purposes, de-identification based on computerized methods is necessary to respect the PHI privacy and ensure its protection from any illegal use or other threatening risks.

VI. Results and Discussion

In this present work, we are interested in studying medical business processes in order to elaborate a data driven clinical pathway model, based on the BPMN language. The aim here is to ensure the respect privacy requirements since early stages of HIS design. Then, we divided clinical data into categories and extracted PHI among them in the form of data model clinical pathway. After that, we defined both personal data protection principles and HIPAA privacy requirements for the specified PHI use and disclosure. Finally, all of the above listed objectives were validated through the modelling of osteosarcoma care pathway business process model chosen as a case study.

As for the completion of the modelling phase of osteosarcoma clinical pathway, we have modeled its complex care pathway which is divided into three phases: *check-up*, *treatment* and *follow-up*. This was done using the actual BPMN language simple patterns. Hence, personal data processing is integrated in the processes, particularly, in a legislation compliant manner which adds more trust to medical documents processing and sharing during the clinical process implementation.

Many difficulties were encountered in clinical pathway modelling using BPMN due to the complexity and multidisciplinary aspect of medical procedures. This has led us to conclude the necessity of a more specialized care pathway modelling language. This has also highlighted the need for a new care pathway modelling and automation language that is sensitive-data driven and could integrate privacy requirements specification. Thus, a new extension of the BPMN modelling language is required.

VII. Conclusion and Future Work

Clinical pathways automation is highly required in standardized HIS. This is traditionally ensured by business process modelling. In this context, we developed a data driven clinical pathway business process model for osteosarcoma, as a case study. We used BPMN as clinical pathway business process modelling language. A shared clinical data model was elaborated further to the clinical business process model.

Since personal data management must obey to data protection law, we defined both personal data protection principles and HIPAA privacy requirements with relation to patients identifying medical documents, in terms of their both use and disclosure.

The adoption of a privacy by design approach offers a better enforcement of privacy since an early stage of computer-based healthcare systems design. This allows an orthogonal integration of privacy obligations throughout the clinical process. For this reason, we are working currently on extending the BPMN process modelling language with privacy annotation features and additional patterns to allow the modelling of privacy specification as part of clinical processes.

We are aiming to define a common vocabulary qualifying clinical pathways specifications with respect to privacy requirements. We believe clinical process modelling languages should be more adapted to a multidisciplinary clinical systems users' profile. Thus, we are planning to investigate the adoption of a variety of symbols and modelling patterns that are better tailored to the requirements of the clinical community.

References

[1] U.S. Department of Health & Human Services HHS: HIPAA Privacy Rule Summary. [Retrieved: July, 2013].

http://www.hhs.gov/hipaa/for-professionals/privacy/laws-regulations/index.html.

[2] E. Rolón et al., "Process modeling of the health sector using BPMN: A case study" The First International Conference on Health Informatics-Diagnostic Pathology (HEALTHINF 2008) IARIA, Jan. 2008, pp. 173-178, ISBN: 978-989-8111-16-6

[3] S. Ferrante, S. Bonacina, G. Pozzi, F. Pinciroli, and S. Marceglia, "A design methodology for medical processes," Applied Clinical Informatics, vol. 7, pp. 191-210, 2016, doi:10.4338/ACI-2015-08-RA-0111.

[4] F. Ruiz et al., "Business process modeling in healthcare," Stud Health Technol Inform, vol. 179, pp. 75-87, 2012.

[5] R. Braun, H. Schlieter, M Burwitz and W. Esswein, "Extending a Business Process Modeling Language for Domain-Specific Adaptation in Healthcare" The 12th International Conference on Wirtschaftsinformatik in Osnabrück-Smart Enterprise Engineering (WI 2015), Mar. 2015, pp. 468-481, ISSN: 0937-6429, ISBN: 978-3-00-049184-9

[6] S. Bielack, D. Carrle, P. G. Casali, and ESMO Guidelines Working Group, "Osteosarcoma: ESMO clinical recommendations for diagnosis, treatment and follow-up," Annals of Oncology, vol. 20, pp. iv137-iv139, May. 2009, doi: 10.1093/annonc/mdp154.

[7] V. Augusto, and X. Xie, "A modeling and simulation framework for health care systems," IEEE Transactions on Systems, Man, and Cybernetics Systems, vol. 44, pp. 30–46, 2014.

[8] E. Rojas, J. Munoz-Gama, M. Sepúlveda, and D. Capurro "Process mining in healthcare: A literature review, "Journal of biomedical informatics, vol. 61, 224-236, June 2016, doi: https://doi.org/10.1016/j.jbi.2016.04.007.

[9] N. Hashemian, and S. S. R. Abidi, "Modeling clinical workflows using business process modeling notation. In Computer-Based Medical Systems," IEEE Computer-Based Medical Systems, pp. 1–4, June 2012 [CBMS 25th International Symposium Italy, 2012].

[10] R. Siegel et al., "Cancer treatment and survivorship statistic," CA: a cancer journal for clinicians, vol. 62, pp. 220-241, Jul-Aug. 2012, doi: 10.3322/caac.21149.

[11] M. S. Isakoff., S. S. Bielack, P. Meltzer, and R. Gorlick, "Osteosarcoma: current treatment and a collaborative pathway to success," Journal of clinical oncology, vol. 33, pp. 3029-3035, Sep. 2015, doi: 10.1200/JCO.2014.59.4895.

[12] A. Luetke, P. A. Meyers, I. Lewis, and H. Juergens, "Osteosarcoma treatment–where do we stand? A state of the art review," Cancer treatment reviews, vol. 40, pp. 523-532, May. 2014, doi: https://doi.org/10.1016/j.ctrv.2013.11.006.

[13] A. Paioli, M. Rocca, L. Cevolani, E. Rimondi, Daniel Vanel, Emanuela Palmerini, Marilena Cesari, A. Longhi, A. M. Eraldo, E. Marchesi, P. Picci and S. Ferrari, "Osteosarcoma follow-up: chest X-ray or computed tomography?," Clinical sarcoma research, vol. 7, Feb. 2017, doi: 10.1186/s13569-017-0067-5.

[14] M. C. Oetzel, and S. Spiekermann, "A systematic methodology for privacy impact assessments: a design science approach," Eur. J. Inf. Syst., vol. 23, pp. 126–150, March. 2014, doi: https://doi.org/10.1057/ejis.2013.18.

[15] S. Rajamani, E. S. Chen, Y. Wang, and G. B. Melton, "Extending the HL7/LOINC Document Ontology Settings of Care," AMIA Annual Symposium Proceedings, Nov. 2001, pp. 994-1001, doi: https://doi.org/10.1016/j.cmpb.2015.09.020.

[16] HIPAA Privacy Rule: Uses and disclosures of protected health information: General rules (§164.502). [Retrieved: January, 2013]. https://www.law.cornell.edu/cfr/text/45/164.502.

[17] HIPAA Privacy Rule: Uses and disclosures for which an authorization is required (§164.508). [Retrieved: January, 2013]. https://www.law.cornell.edu/cfr/text/45/164.508.

[18] Regulation (eu) 2016/679 of the European parliament and of the council of 27 April 2016 on the protection of natural persons with regard to the processing of personal data and on the free movement of such data, and repealing directive 95/46/ec (general data protection regulation). [Retrieved: March, 2017]. https://eur-lex.europa.eu/legal-content/EN/TXT/?uri=uriserv%3AOJ.L_.2016.119.01.0001.01.ENG&toc=OJ%3AL%3A2016%3A119%3ATOC.

A Cost-effective BCI Assisted Technology Framework for Neurorehabilitation

M A Hannan Bin Azhar[1], Anthony Casey[1] and Mohamed Sakel[2]
[1] Computing, Digital Forensics and Cybersecurity, Canterbury Christ Church University, UK
[2] East Kent Hospitals University, NHS Foundation Trust, Canterbury, UK
Email: hannan.azhar@canterbury.ac.uk; antcasey07@outlook.com; msakel@nhs.net

Abstract -- **Brain Computer Interface (BCI) controlled assistive robotic systems have been developed with increasing success with the aim to rehabilitate brain injured patients to increase their independence and quality of life. While such systems may use surgically implanted sensors, non-invasive alternatives can be better suited due to ease of use, reduced cost, improvements in accuracy and reliability with the advancement of the technology and practicality of use. The consumer grade BCI devices are often capable of integrating multiple types of signals, including Electroencephalogram (EEG) and Electromyogram (EMG), as well as basic motion-based signals, such as gyroscopic data. This paper reports the development of a framework for rolling out cost-effective BCI driven assistive technology systems and details the implementation and evaluation of a prototype robotic system to determine the efficacy of the proposed framework. The results indicate that the first stage of the framework was effective in accuracy, safety, usability, portability, adaptability and personalisation.**

Keywords* - *BCI; Assistive Technology; EEG; EMG; Disability; Neurorehabilitation.

I. INTRODUCTION

There are over 12.5 million people in the UK living with a neurological condition, out of which over a million are substantially disabled by their condition and a further 350,000 require help for most of their daily activities [1]. These neurological conditions cost the National Health Service in the UK (NHS) £4.4 billion in 2014 [1] (4.2% of the NHS expenditure), with the cost set to treble over the next 20 years [2]. An estimated £2.4 billion is spent on social care services alone for people with a neurological condition [1]. Given the wide range of issues that neurorehabilitation covers, as well as the success in the implementation of Brain Computer Interface (BCI) based systems using both Electroencephalogram (EEG) and Electromyogram (EMG) signals [3]-[11], it is apparent that BCI based assistive systems could become viable alternatives to traditional neurorehabilitation methods.

There is limited evidence of BCI assisted neurorehabilitation available in the literature for post-stroke disorders [3], paraplegia [4], spinal cord injury [5] and tetraplegia [6]. Cesqui et al. [10] reported an EMG based robotic system to assist in the rehabilitation of patients and Soekadar et al. [11] investigated the use of EMG as a method of post-stroke rehabilitation. However, success of such technology assisted rehabilitation largely relies on the effectiveness of integration through a process defined in a neurorehabilitation framework that is more personalized and convenient [12][13]. Development and implementation of an assistive technology has to be carefully phased out using rigorous procedures described in a framework for a system [14][15] to be better suited for patients. Availability of a framework specific to patients with more personalised needs and requirements of bespoke technologies were discussed; especially the need for a framework for the integration of BCI within neurorehabilitation context was mentioned in [5]. As a solution, non-invasive off-the-shelf portable BCI assisted systems have been popular choices due to improvement in accuracy and reliability of such systems with the advancement of technology and practicality of use [16]. This paper proposes a cost-effective BCI based assistive technology framework for the rehabilitation of patients suffering from neurological disabilities, which is an adaptation and merge of two frameworks: one [12] details the entire rehabilitation process and the other [14] details the technology lifecycle (development, selection, learning and integration). In addition to developing a framework for neurorehabilitation, this paper also investigates the efficacy of the proposed framework by developing a prototype robotic system with the help of two portable devices: an Interaxon Muse BCI (to gather brain signals) and a UFactory uArm (for robotic control), as shown in Figure 1.

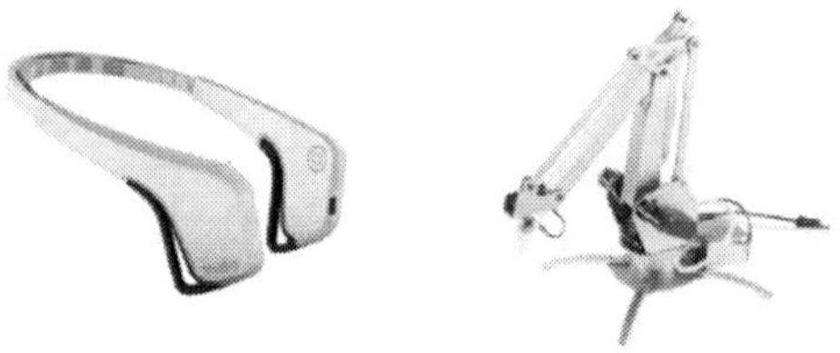

Figure 1. Interaxon Muse (left) and UFactory uArm (right).

The remainder of this paper is organised as follows: Section 2 gives an overview of a proposed framework, Section 3 details the evaluation of the framework using a prototype system and its implementation. Section 4 describes the experimental scenarios to evaluate the framework. Results will be discussed in Section 5. Finally, Section 6 concludes the paper.

II. PROPOSED FRAMEWORK

We propose a new framework for BCI assisted neurorehabilitation. We consider three distinct stages of the framework lifecycle: system development, clinical trials and operational. It is critical that the end product of this system is reviewed by one or more specialists to determine whether the system has sufficient accuracy, safety, usability, portability, adaptability and personalisation.

A. System Development

The system development stage consists of four actors, adapted from Kintsch [14]; a trial group, one or more trial caregivers, a specialist and one or more developers (as listed in Table 1). While Kintsch [14] specified the required traits for each actor, this paper specifically states the role of each actor within the system.

TABLE I. ROLES OF ACTORS IN SYSTEM DEVELOPMENT STAGE

Trial Group	Trial Caregiver	Specialist	Developer
Volunteers who trial the system. Require no prior training on the system. Gives feedback to specialist on usability of the system.	Instructs the trial user on how to operate the system. Initialises any initial parameters of the system and fits the BCI. Requires minimal training in the system.	Reviews data given by the system. Gives feedback to the developer based on feedback from system and user.	Makes changes to the system based on feedback. Ensures system safety and good coding conventions for adaptability.

This stage focusses primarily on the planning and development of the system to test the feasibility. The main criteria for suitability are the relevance to the neurorehabilitation task at hand, the cost-effectiveness for easy prototyping and overall safety of the system. If the system is not suitable, it is the specialist's decision whether further development should proceed or not. Figure 2 shows a sequence diagram of the system development stage. Once the initial proposal is being approved, the developer should strive to make the code as modular as possible to ensure that the code is easy to adapt and maintain in the future. Members of the trial group will run through the calibration and experimentation to gather data on system efficacy. Feedback from both the specialists and users is also critical to evaluate the suitability of the system in order to take it forward to the next clinical trial stage.

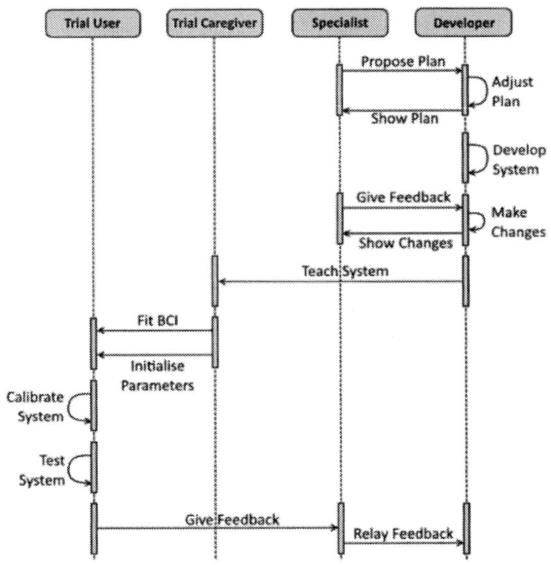

Figure 2. Sequence diagram of the system development stage.

Since the trial caregiver will be responsible for both fitting the BCI to the user's head as well as setting the experimental parameters, it is vital that the caregiver (if they are not also the developer) receives some form of training from the developer prior to carrying out any experiments. Users should carry out calibration at least once before running any experiment to filter out any unwanted artefacts. Calibration can either be run every time the experiments are to be carried out, or the data can be stored in a user profile and loaded each time the experiment is run.

B. Clinical Trials

Before the clinical trials stage commences, the ethical approval must be completed to try out the systems on real patients. A select group of neurologically disabled patients will volunteer to trial the system and give feedback on the system. The developer will act on this feedback and the advice of the specialist. Caregivers are likely to be nurses or family members of the patients. Table 2 shows the roles of all actors in this stage. The specialist will now be in a position to handle the training of the system to the caregivers. This is likely to be done through group training sessions where specialists train nurses. Figure 3 shows the sequence diagram of the clinical trials stage.

TABLE II. ROLES OF ACTORS IN CLINICAL TRIALS STAGE

Patients	Caregiver	Specialist	Developer
Neurologically disabled volunteers that trial the system. Require no prior training on the system. Gives feedback to specialist on usability of the system.	Either nurses or family / friends of the patients. Instructs the patient on how to operate the system. Initialises any initial parameters of the system and fits the BCI. Requires minimal training in the system.	Reviews data given by the system. Peer review through conferences in hospitals and Patient Association Forums. Gives feedback to the developer based on feedback from system and user.	Makes changes to the system based on feedback. Ensures system safety and good coding conventions for adaptability.

C. Operational Stage

The operational stage is the final stage of the framework, which involves regular maintenance of the system to ensure that the technology is being kept up-to-date and patched against bugs. Table 3 shows the roles of all actors in the final stage of the framework. The system at this stage is ideally to be phased out to wider group of patients who wish to make use of it. The specialist can use the data gathered from patients by the system to evaluate how well the rehabilitation process is going on. This evaluation can then be shared with the developer and any pertinent improvements can be justified through this process, as illustrated in the sequence diagram in Figure 4.

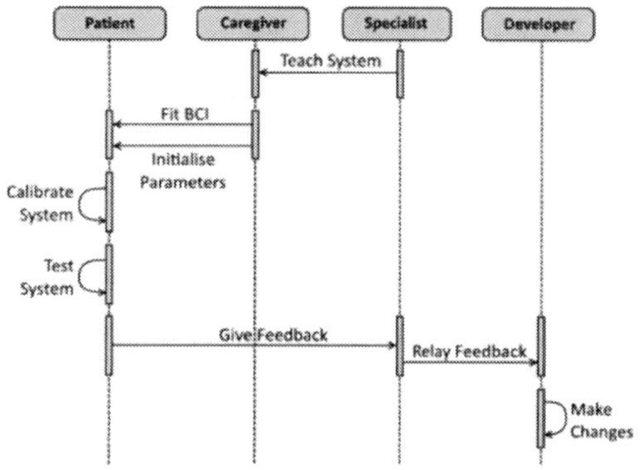

Figure 3. Sequence diagram of the clinical trials stage.

TABLE III. ROLES OF ACTORS IN OPERATIONAL STAGE

Patients	Caregiver	Specialist	Developer
Neurologically disabled patients that use the system. Require no prior training on the system.	Either nurses or family / friends of the patients. Instructs the patient on how to operate the system. Initialises any initial parameters of the system and fits the BCI. Requires minimal training in the system.	Reviews data given by the system.	Makes changes to the system as needed. Ensures system safety and good coding conventions for adaptability.

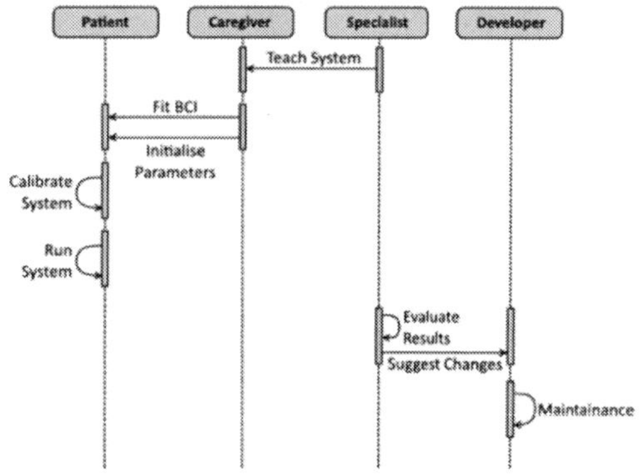

Figure 4. Sequence diagram of the operational stage.

III. FRAMEWORK EVALUATION

This section details a proof of concept of the framework proposed in Section 2, consisting of the operation of a robotic arm (as the assistive technology component) using a Muse BCI device (as the BCI component). Given the time constraints on this project, the evaluation was carried out only for the stage 1 of the framework (Figure 2). Actors involved in this stage were a developer (also acting as a trial caregiver), a specialist and a small trial group of 5 subjects. Accuracy and usability (as required by the framework) measures will be observable from the experimental results of the system, whereas the other performance measures will be discussed within the evaluation.

A. *System Overview*

The system hardware consists of a PC, a small portable BCI device, Muse and a portable desktop robotic arm, uArm. Viability of such low-cost portable devices as a solution to neuro-rehabilitation was discussed with a specialist at the Kent and Canterbury NHS hospital's neurorehabilitation department. Using rigorous use case scenarios both developer and specialist agreed on achievable requirement specifications for the prototype system. The overall architecture of the system is illustrated in Figure 5 and consists of two main branches: the EMG data branch and the gyroscopic data branch. More details of these steps are shown in the figure.

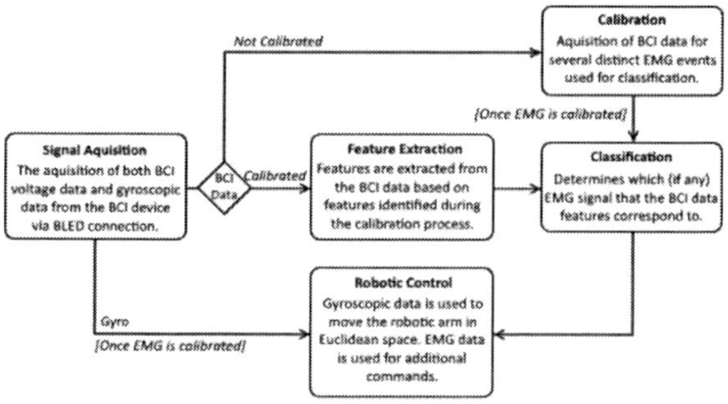

Figure 5. Overview of the system data flow.

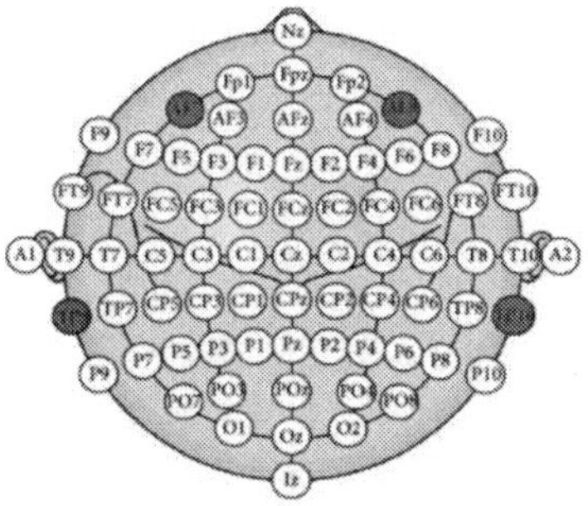

Figure 6. Muse Electrode Placement modified from [17].

B. *Data Acquisition*

The first step is to obtain data from the Muse BCI device, which is achieved via hosting of a Python server leveraging the benefits of a Bluetooth Low-Energy Dongle (BLED) connection direct to the Muse device. Several sets of data are supported by this server, but the most notable are the four electrodes; AF7, AF8, TP9 and TP10 (using the 10-20 international system seen in Figure 6) at a sample rate of 256Hz as well as the gyroscopic data (at a sample rate of ~50Hz). These electrodes are able to detect electrical signals across the front of the skull, namely EMG artefacts such as blinks, winks and jaw clenches.

C. Calibration

Calibration covers the personalisation of the system, by creating separate user profiles containing information specific to each user. The user is presented with a series of prompts over the calibration process and is given a set of ten (temporally equidistant) prompts telling them to rest, which gets the baseline brain activity level. After the tenth prompt (to get a reliable amount of data), the user is given another ten prompts instructing them to blink, which gets the voltage associated with the blinking. Likewise, there is another ten prompts for left wink and then another ten for the right wink.

As an example, Figure 7 shows EMG readings over all electrodes for two different EMG artefacts (left and right winks). The figure shows that during left winks, the spike in voltage for the left electrodes are greater than the spike in voltage for the right electrodes. When there is a right wink, the spike in voltage for the right electrodes are larger than the voltage spike for the left electrodes.

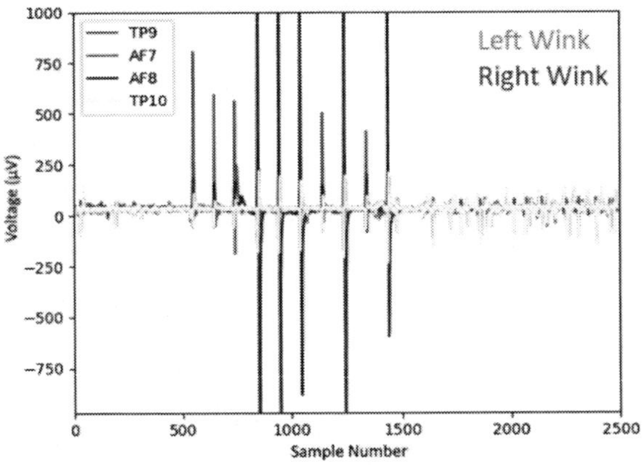

Figure 7. Differentiating between left and right wink using voltages.

Once all 40 prompts are finished, an EMG profile is built for the user. This profile consists a single threshold value, as shown in (1), calculated from the average of the maximum and average of the mean EMG values over the samples, $n=10$ in (1), for every user for each of the different sets of BCI electrodes during blinks, left winks and right winks.

$$Threshold = \left(\frac{1}{n}\sum_{i=1}^{n} Vmax_i - \left(\frac{1}{n}\sum_{i=1}^{n}(Vmax_i - Vavg_i)\right) \times tolerance\right) \quad (1)$$

A tolerance value was set empirically to configure a better responsive system in relation to variation of voltage levels corresponding to EMG spikes. This algorithm ensures that a more responsive and personalised profile is set for each individual user.

D. Classification

Classification deals with the accuracy of the system. A successful and responsive classifier will yield high levels of accuracy. The classification step involves the comparison of EMG voltage data in real-time with the EMG voltage data stored in a given user's profile and applying semantics to that signal, i.e., to map a specific voltage combination to a particular EMG signal for that user.

The calibration file contains a value (which acts as a threshold) for each electrode during blinks, left winks and right winks. If both the left and right electrodes pass their respective electrode values in the calibration file, then the signal is classified as a blink. If the real-time left electrode values are greater than the left electrode's data in the calibration file and the right electrode value is less than its respective calibration value, then the signal is classified as a left wink. Likewise, if the right electrode values are greater than the calibration value and the left electrode value is less than its respective calibration value, the signal is classified as a right wink.

If none of the above criteria are fulfilled, the system returns that there was no significant EMG signal detected and therefore, no action is given to the uArm to avoid carrying out any unexpected actions by the arm. For clarity, the output of each classification is also reported back to the user.

E. Robotic Control

After calibration, the gyroscopic data can be streamed from the Muse and used to control the robotic arm, where the user moves their head in Euclidean space which corresponds to the three degrees of freedom in the robotic arm (X, Y and Z axes). The robotic arm can only move within a pre-defined "bounding box"; while this is set as an experimental parameter, is a major step to ensure system safety. Simultaneously, voltage data is being streamed from the Muse, which is used to perform a variety of actions on the uArm. Each time the program is run, it will allow users to dynamically load a control protocol, meaning that custom systems can be reused over several different experiments, or control protocols that are optimised for a specific use can be loaded.

IV. EXPERIMENTAL SCENARIO

Once the system has been sufficiently developed as described in Section 3, the next step in the framework is to teach trial caregivers how to fit the BCI and initialise the experiment parameters, system calibration and the acquisition of experimental data. After receiving feedback on the system, changes are made (if necessary). The feedback will determine the accuracy and usability of the system.

An experiment was designed to determine how usable the system is in a real-world scenario; in this case, an experiment with feeding task was set upon discussion with a specialist as a feasible task for neuro-rehabilitation. The task involves users to attempt to use EMG signals and head movements (gyroscopic data) to move the uArm (using a spoon as an "end effector") to a plate of dummy food (using paper balls), scoop up some of the "food" and to move it to a predefined mouth level. The process involves a user moving the uArm first using gyroscope movements and then scooping up some food. The user should then blink, which brings the food up to a predefined mouth level automatically. Once the food is being "eaten" or removed, another blink brings the uArm back down to the plate level automatically. After the trial of this task, the user should then repeat the steps for a total of three times to ensure the sufficient use of the system to be able to comment on the efficacy of the system.

Figure 8 demonstrates the control protocol of the uArm during the feeding experiment. In free mode (the initial state of the system), the subject can use gyroscopic events to move the robotic arm in the Euclidean space. If the user blinks while the uArm is not at the mouth level, the uArm moves to the mouth level and locks itself, preventing further movement. Blinking while the uArm is at the mouth moves the uArm back to the plate level and unlocks it, meaning that the user can use gyroscopic data to move the uArm again. A right wink will result in resetting the position of the uArm to its default starting position.

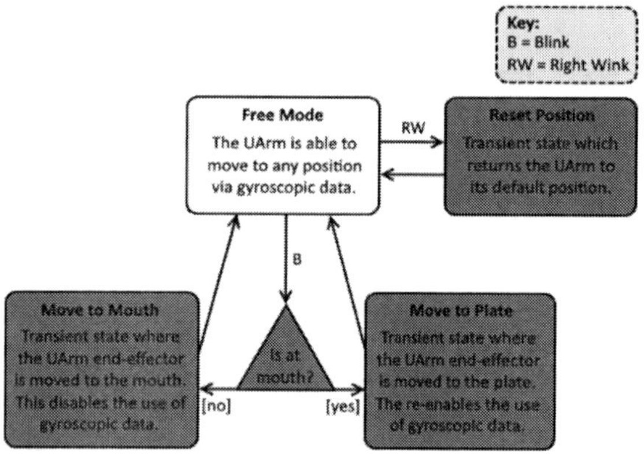

Figure 8. EMG control protocol for the robotic arm during the feeding experiment.

The objective metrics for success of this experiment are the time taken to bring the food to the user's mouth, whether the uArm successfully traversed from the plate to the mouth, the number of items dropped in this process and finally, whether the uArm successfully traversed from the user's mouth back to the plate. In addition to the objective metrics, subjective measures shall also be taken in the form of a short survey of user experience, following the system usability scale (SUS) [18] comprising of the ten questions answered on a scale of one (strongly disagree) to five (strongly agree).

V. Results of Evaluation

This section reports the work in progress results of the evaluation of the framework's development phase using the feeding experiment as detailed in Section 4. At the time of writing this paper, in total 15 trials (Table 4) were completed to evaluate the success of the feeding task in a non-clinical setting using a trail group of five able-bodied volunteers contributed to three trials each. Performance measures of the system and the usability results were discussed in detail with a specialist at the neuro-rehabilitation centre to gain specialist's critiques on feasibility of the system if it was to be tried on real patients during the clinical trial phase of the framework.

The results of Table 4 shows that users were able to adapt to the system at varying speed. Trial times to complete the route from plate to the mouth level varied amongst users from around 7 seconds to more than 3 minutes. Even though 3 minutes seems to be longer time to complete the task, according to specialist's comments it was still a good outcome as hiring nurses or carer to carry out the same task would cost more for rehabilitation. The mean time for completion of the same task was ~45.86 seconds, with a standard deviation of ~46.55, however for one trial it took unexpectedly long (194.62 seconds) to complete the task. Discarding the outlier of 194.62 seconds, the mean completion time reduces to 35.24 seconds with a standard deviation of ~25.06.

TABLE IV. Results of the Feeding Experiment

		Plate to Mouth			Mouth to Plate
ID	*Trial*	*Time (sec)*	*Route Complete (Y/N)*	*Items Dropped*	*Route Complete (Y/N)*
1	1	11.51	Y	0	Y
1	2	45.49	Y	0	Y
1	3	7.20	Y	0	Y
2	1	8.32	Y	0	Y
2	2	13.09	Y	0	Y
2	3	8.80	Y	0	Y
3	1	34.30	Y	0	Y
3	2	25.36	Y	2	Y
3	3	194.62	Y	2	Y
4	1	72.58	Y	0	Y
4	2	14.82	Y	0	Y
4	3	68.29	Y	0	Y
5	1	77.00	Y	0	Y
5	2	56.97	Y	0	Y
5	3	49.47	Y	0	Y

The system managed to successfully navigate from the plate level to mouth level and vice versa in every single trial and the specialist found this to be a reliable outcome. Number of items dropped while completing the moving task from plate to the mouth level also varied amongst users depending on their efficacy of using the system after the training session. Except one user, no one dropped any item in any of the trails while moving the arm to the mouth level, showing an acceptable level of efficacy. For one user, blink event did not produce enough electrical potential to be registered as an EMG event. During the calibration stage, after checking all possible EMG events which can be detected by the system, it was found that "frowning" generates higher electrical potential for this user instead of the blink act. So, during the training session, instead of the blink act frown act was registered as an EMG control input for this user. This shows adaptation of the system control input based on specific user physiological requirements, supporting the adaptability requirement of the framework.

TABLE V. SUS SURVEY RESULTS

Question	Average Score
I think that I would like to use this system frequently.	2.6
I found the system unnecessarily complex.	2.2
I thought the system was easy to use.	4
I think that I would need the support of a technical person to be able to use this system.	1.8
I found the various functions in this system well integrated.	3.4
I thought there was too much inconsistency in this system.	2.6
I would imagine that most people would learn to use this system very quickly.	4
I found the system very cumbersome to use.	2.6
I felt very confident using the system.	4
I needed to learn a lot of things before I could get going with this system.	2

Table 5 summaries the results of the SUS survey which was completed by users at the end of their trials. The results indicate that users found the system easy to use and they also felt confident using the system. There was strong agreement amongst users about easily learning to use the system and they also believed that most people would learn to use the system very quickly. Some users found the system cumbersome to use and inconsistent in completing the task, which was also the case in objective measures reported in Table 4, as the users completed the task in largely varying times.

VI. CONCLUSION

This paper proposed a new BCI driven assistive technology neurorehabilitation framework and reported the work in progress evaluation of the development stage of the framework. Specialist involvement in design and evaluation of the framework gave valuable insight into successful adaptation of the technology to fit better for patients' care for neurorehabilitation. The prototype system suited well within the framework with some success as observable from the results of the evaluation. While the system was deemed acceptable by the specialist, some changes to the system need to be addressed; such as more rigorous trials are required to be carried out on a larger sample population consisting of a wider variety of subjects accounting for differences in age, gender and abilities. Latency in the system should also be improved. Further experiments over a greater length of time could determine whether users can improve their performances over a number of trials. Discussion has been made already to use the framework to develop more complicated BCI assisted system, such as the BCI controlled exoskeleton, for the purpose of neurorehabilitation.

REFERENCES

[1] The Neurological Alliance, "Neuro numbers (Article)", April 2014. Available from: http://www.neural.org.uk/store/assets/files/381/original/Final_-_Neuro_Numbers_30_April_2014_.pdf [retrieved: September, 2018].

[2] Stroke Association, "Current, future and avoidable costs of stroke in the UK (Article)", September 2017. Available from: https://www.stroke.org.uk/sites/default/files/costs_of_stroke_in_the_uk_report_-_executive_summary_part_1.pdf [retrieved: September, 2018].

[3] L.E. van Dokkum, T. Ward, and I. Laffont, "Brain computer interfaces for neurorehabilitation – its current status as a rehabilitation strategy post-stroke", Annals of Physical and Rehabilitation Medicine, vol. 58, no. 1, pp. 3-8, 2015.

[4] C.E. King et al, "The feasibility of a brain-computer interface functional electrical stimulation system for the restoration of overground walking after paraplegia", Journal of Neuroengineering and Rehabilitation, August 2015.

[5] B.C.A. Osuagwu, "Neurorehabilitation of hand functions using brain-computer interface", PhD thesis, University of Glasgow, September 2015.

[6] D.C. Irimia, M.S. Poboroniuc, F. Serea, A. Baciu, and R. Olaru, "Controlling a FES-EXOSKELETON rehabilitation system by means of brain-computer interface", International Conference and Expositionon Electrical and Power Engineering (EPE), October 2016.

[7] J.J. Daly and J.R. Wolpaw, "Brain-computer interfaces in neurological rehabilitation", The Lancet Neurology, vol. 7, no. 11, pp. 1032-1043, November 2008.

[8] C.S. Ang, M. Sakel, M. Pepper, and M. Phillips, "Use of brain-computer interfaces in neurological rehabilitation", British Journal of Neuroscience Nursing, vol. 3, no. 3, pp. 523-528, September 2013.

[9] N. Birbaumer, "Breaking the silence: brain-computer interfaces (BCI) for communication and motor control", Psychophysiology, vol. 43, no. 6, pp. 517-532, November 2006.

[10] B. Cesqui, P. Tropea, S. Micera, and H.I. Krebs, "EMG-based pattern recognition approach in post stroke robot-aided rehabilitation: a feasibilty study", Journal of NeuroEngineering and Rehabilitation, July 2013.

[11] S.R. Soekadar, N. Birbaumer, M.W. Slutzky, and L.G. Cohen, "Brain-machine interfaces in neurorehabilitation of stroke", Neurobiology of Disease, vol. 83, pp. 172-179, November 2015.

[12] British Society of Rehabilitation Medicine, "Specialist neurorehabilitation services: providing for patients with complex rehabilitation needs", Specialised Neurorehabilitation Service Standards, April 2015.

[13] X. Feng and J.M. Winters, "An interactive framework for personalized computer-assisted neurorehabilitation", IEEE Transactions on Information Technology in Biomedicine, vol. 11, no. 5, September 2007.

[14] A. Kintsch and R. DePaula, "A framework for the adoption of assistive technology", Supporting Learning Through Assistive Technology (SWAAAC), pp.1-10, January 2002.

[15] Ministry of Education, "Assistive technology – assessment framework", January 2014. Available from: http://inclusive.tki.org.nz/assets/inclusive-education/resource-documents/AT+Fact+Sheet+Assessment+Frame+Jan+2014-2.pdf [retrieved: September, 2018].

[16] F. Deravi, C.S. Ang, M.A.H.B. Azhar, A. Al-Wabil, M. Phillips, and M. Sakel, "Usability and performance measure of a consumer-grade brain computer interface system for environmental control by neurological patients", International Journal of Engineering and Technology Innovation, vol 5, pp.165-177, July 2015.

[17] F. Sharbrough, G.E. Chatrian, R. Lesser, and T.W. Picton, "American electroencephalographic society guidelines for standard electrode position nomenclature", Journal of Clinical Neurophysiology, vol. 8, no. 2, pp. 200–202, 1991.

[18] J. Brooke, "SUS – A Quick and Dirty Usability Scale", 1986. Available from: https://hell.meiert.org/core/pdf/sus.pdf [retrieved: September, 2018].

Blessed Health - A Journey From "I-Health" to "We-Health"

Srikali Goutam Varanasi
Data Analytics Consultant, Kaiser Permanente
Oakland, USA
srikalig@yahoo.com

***Abstract*— Health is a fundamental feature of human beings. Health is not just an interaction with health providers in order to arrive at a diagnosis and a treatment process. Health is a summation of several elements, including life style, food habits, confidence, and emotional interactions with others. While the traditional I-Health (health of an isolated individual) model has its limitations, We-Health (Community based informed health) has tremendous potential due to its strength and spread. The thought of one's unhealthiness leads to physical and psychological detriment. Individuals throughout the world have varied heights, complexions, personalities, beliefs, and diverse ways of living. In the same way, their health and vital numbers are relative to their personal, physical, psychological, and environmental characteristics. Blessed health is an initiative to bring the awareness of 'Informed Health', referring to the relative health of individuals.**

Keywords – Blessed; Informed; Health; Community; We-Health.

I. INTRODUCTION

Blessed health is a dynamic model of health, where the health of communities is targeted, instead of individualized health assessments. Health becomes a dynamic practice in this "We-Health" model, compared to that of a static "I-Health" model. This new paradigm will effectively combat the current issues of depression, mental health, drug dependency, drug addiction, and disease management solely by lifelong medications.

II. CURRENT SCENARIO- "I HEALTH"

The current health scenario is a static model, where individuals get their numbers checked up. These numbers include blood pressure, cholesterol, and other values, which are then compared against national averages and standards. Individual treatments and prescriptions proceed, resulting in the perceived health of an individual. In most cases, health care providers are under pressure to map the deviations in the numbers to some health condition, resulting in a mandatory prescription of medications. The medications may result in adverse drug reactions [1].

III. "I-HEALTH" PARADIGM

The current "I-Health" paradigm depicted in Fig 1, can be explained as follows:

1. Input – The paradigm begins with an individual's health check-up and recording of associated numbers, such as vitals, counts, etc.
2. Process – Next, numbers are compared to the "standard" or "average" data. This leads to stress and fear in the patient of possible health conditions. Then, "common" or "standard" procedures/treatments ensue.
3. Output/ Result – "I-Health" which is the perceived health of an individual, resulting in a lack of hope in becoming disease-free within their life time. Individuals feel pressured to continue managing diseases or medical conditions solely through prescription drugs.

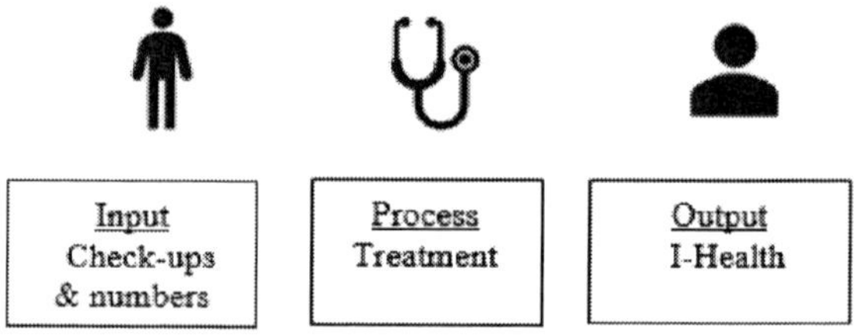

Figure 1. I-Health Paradigm

IV. CHALLENGES IN "I-HEALTH"

Comparison of an individual's health numbers to standards will result in erroneous conclusions, since the comparison is not holistic. An individual's medical condition is assessed by symptoms, causes, medical history, and hereditary aspects at the time of assessment. This will not take into consideration an individual's physical state, psychological state, lifestyle, and environment. This kind of assessment will result in a diagnosis, based on an approximation, due to the deviation from the "norm". Sometimes, in this model, individuals aren't fully aware of all the possible treatment options and therefore obligated to continue the ongoing treatment.

V. WHY "BLESSED HEALTH"- A DYNAMIC MODEL IS PREFERRED

Health is a universal cause. Health is everyone's responsibility and a fundamental right. Health is not just the right of only affluent populations. It is a community effort [2]. Though some work has been done in this area, most of it relates to communication aspects of health rather than community-based health [3]. This community-based health is a preferred dynamic model, called "Blessed Health" which is a paradigm shift from "I-Health", and is a journey towards "We-Health".

VI. "WE-HEALTH" SCENARIO

The proposed "We-Health" paradigm depicted in Fig 2 can be explained as follows:

1. Input – In this model, individual's health assessment is mapped to the health conditions of the individual. Here, symptoms are approached in a holistic way considering the physical, psychological, lifestyle and environmental factors. Informed health conditions of family, community and surroundings help to interpret health conditions in the right way without fear of numbers. Here every effort is made to avoid intrusions, negative interpretations, and unnecessary engagement. Positivity about individual's current health condition is a natural blessing here.
2. Process – Health is nature's gift. Inclination towards health does not cause fear or stress and it creates a strength climate. Strength brings confidence. As a result, one will naturally be motivated towards spreading the positive health around.
3. Output/ Result – The result of a cohesive process is a healthy community consisting of the individual, one's family, extended family, friends, and the whole surrounding ecosystem. The "Spread" includes nurturing, mentoring and adapting ways, to analyze and improve health of everyone in the ecosystem.

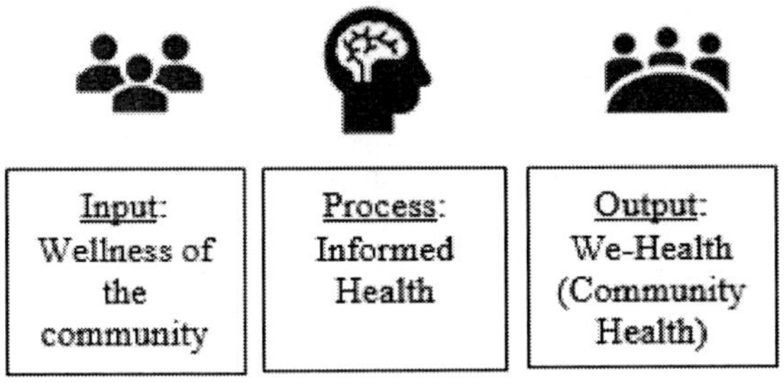

Figure 2. We-Health Paradigm

Viewing health as a dynamic model promotes operational health. In this model, health is everyone's moral and practical responsibility, that is shared amongst communities. A simple protocol can be adopted to achieve this.

1. Owning one's health.
2. Being accountable.
3. Being mindful.
4. Knowing one's food.
5. Know how food comes to one's plate. One's food plate should not be a detriment to another life in the ecosystem [4].
6. Engage and motivate community.
7. Educate self and surroundings.
8. Live cohesively with the environment to create a healthy ecosystem.

VII. ANCIENT SUCCESSFUL MODELS OF COMMUNITY HEALTH (WE-HEATH)

There are numerous references in Ayurveda, a system of medicine practiced by ancient Indians emphasizing the importance of community-based health.

An excerpt from these ancient texts, refers to health as collaborative effort of teamwork, strength, blessing, and peace, as described below: "One shall strive to protect one's own body, mind, and energies. Then one shall protect and care for their families, communities and surroundings. Then one shall protect the totality of ecosystem".

In the ancient Sanskrit texts, health is perceived as the combination of 5 layers of health of an Individual. The medicines that one takes into the body only can cure one-fifth of the totality of health, which is at the physical level. To understand a symptom and to have blessed health, one should work on acquiring all the 5 layers of health.

The five layers of health are described as follows: Every human being has 5 sheaths or layers for blessed health that can contribute to total or holistic health. Those layers are:

1. "Physical" health, which is related to the physical body nourishing with food and water.
2. "Breathing" health, which is the life force related to breathing and stress preventing techniques.
3. "Mind" health, which cultivates the inner and outer energies related mental and emotional aspects.
4. "Intellectual" health, which is related to one's own intellect, through logic, decisioning, thinking, interacting, and balancing.
5. "Blessed" health, which is a totality of health where one is ready to extend help to others in the community and achieve strength and spread.

These five layers of health achieved with the following simple protocol:

1. Eat plant-based food, feel well, and work well.
2. Control your breath and enhance vital energies with good habits and work ethics, which in turn reduces stress.
3. Clear your mind with meditation and positive interactions to channel mental and emotional energies for positive health and progress.
4. Acquire knowledge, interact with community, and make informed decisions with mindfulness.
5. Become a vital leader in spreading health and happiness.

VIII. Conclusion

Health in general is perceived and understood like an ancient story of elephant and 5 blind men. Once five blind men approached an elephant for the first time and wanted to figure out what "the object" is. Each blind man touched a different part of the elephant and perceived it differently. The first one thought it to be a rope by touching the tail. The second one perceived it to be a mountain by touching the body of the elephant. The next one thought it to be a stick by feeling the tusk. Yet another one thought it to be a pillar, and so on. This is known as the theory of blindfold predictions. All of them may have different perceptions, but the truth is beyond perceptions. In a comparable way, one should understand the concept of health in its totality and essence. Assessing health at physical level only can contribute to one fifth of the total of health.

In "We-Health" model, health is a summed-up equation of health at all levels including the physical, life-force, mind, intellectual, and interactional. All of it adds up to the totality of health referred to in this article as "We-Health" or "Blessed Health". In this model, there is no room for depression and hence other prevalent common mental health issues can be prevented.

Acknowledgment

I thank Sudhindra Goutam Vedanthi for mentoring and being supportive throughout the development of this idea. I thank Srivishnu Goutam Vedanthi and Sumedh Goutam Vedanthi for proof reading and editing.

References

[1] G. Null, C. Dean, M. Feldman, D. Rasio, and D. Smith, "Death by Medicine", 2004.

[2] D. Blumenthal and R.DiClemente, Community-Based Health Research: Issues and Methods. New York: Springer Publishing Company. October 2003. 27 ISBN: 0-8261-2025-3.

[3] R. Schiavo, Editor-In-Chief, "Importance of community-based communication for health and social change", Journal of Communication in Healthcare, Vol 9, No. 1, 2016

[4] W. Harris, The scientific basis of vegetarianism, Hawaii Health Publishers, 1995, ISBN-13: 978-0964653801.

Chronic Exposure to Chlorpyrifos Alters Spontaneous Motor Activity and Skeletal Muscles Contractility in Young Rats

Nancy Hallal, Mohamad Sayed,
Department of Biological Sciences,
Faculty of Sciences, Beirut Arab University,
Lebanon.
E-mail: nancy.hallal@liu.edu.lb
Department of Biochemistry,
Faculty of Sciences, Alexandria University,
Egypt
E-mail: mohamed.moustafa@bau.edu.lb

Wiam Ramadan, Hassan Khachfe, Ali Salami,
Maram Saleh, Hiba El Sabbouri, Wissam H. Joumaa,
Lebanese Institute for Biomedical Research and Application (LIBRA),
International University of Beirut (BIU),
Beirut, Lebanon
E-mails: wiam.ramadan@liu.edu.lb,
hassan.khachfe@liu.edu.lb,
Université Libanaise, Faculté des Sciences,
Laboratoire de Physio-Toxicité Environnemental,
EDST, ER 017, Nabatieh, Lebanon
E-mails: ali.salameh@liu.edu.lb,
maram.saleh@hotmail.com, hiba.sabboury@outlook.com,
wjoumaa@ul.edu.lb

***Abstract*—Chlorpyrifos (CPF) is one of the most common Organophosphorus Pesticides (OP) used worldwide. The toxicity of CPF results from the phosphorylation of the acetylcholinesterase enzyme at nerve endings leading to a cholinergic poisoning in both central and peripheral nervous system. Neurobehavioral alterations, such as cognitive deficits and locomotor impairment, have been extensively reported in studies following an acute exposure to CPF, but to less extent following a chronic exposure. The aim of this study is to examine the effects of chronic dietary exposure to CPF for 6 weeks on the spontaneous motor activity and the physiological parameters of two skeletal muscles, soleus and *extensor digitorum longus (edl)*, involved in motor activities in young rats. Young rats were fed diets containing 1 mg/kg/day CPF (CPF1) or 5 mg/kg/day CPF (CPF5) for six weeks. Behavioral testing to assess the locomotor activity has been performed on weekly basis. The fibers from soleus and *edl* muscles were dissected and used to study contractile properties. Our results showed that animals treated with CPF suffered from hypolocomotion dose independent, which have been clearly manifested as an increase in latency time as well as the number of errors assessed by the beam walking test and a significant decrease in the beam balance time. CPF5 exposure showed an increase in twitch tension comparing to control groups for both soleus and edl. However, CPF1 exposure induces a decrease in twitch contraction in edl muscle. Repeated low level exposure to CPF1 and CPF5 impaired the fatigability index of soleus but not *edl* muscle. Those dose independent decrements reported in the behavioral tests came along a significant effect on the muscle's contractile parameters. Thus, the greater level of contraction in both soleus and edl muscles studied in vitro might be linked to the alterations shown in the locomotor activity of rats.**

Keywords-pesticides chlorpyrifos; spontaneous motor activity; muscle contractility.

I. INTRODUCTION

Even though lots of governmental regulations are being engaged to limit the market growth of chlorpyrifos (CPF) over the next seven years (up to 2022), it remains one of the most common Organophosphorus Pesticides (OP) used worldwide [1]. Its toxicity results primarily from the inhibition of the AcetylCholinesterase Enzyme (AChE) at nerve endings that will lead to the accumulation of the acetylcholine (Ach) in the synaptic cleft. This will overstimulate the effector organ, such as neurons in the Central Nervous System (CNS) and the skeletal muscle cells and will lead to cholinergic poisoning [2]. Neurobehavioral alterations, such as cognitive deficits and locomotor impairment, have been extensively reported in previous studies following an acute exposure to CPF via two main routes subcutaneous [3] [4] or intragastric administration [5], but to less extent following a chronic exposure via one of the main route "the ingestion of contaminated food". The aim of this study is to investigate whether the repeated exposure of young rats for six weeks to low doses CPF will affect 1) the spontaneous motor activity assessed by the beam walking test and prehensile traction test and 2) the physiological parameters of two skeletal muscles involved in the motor activities: the soleus, a slow-oxidative muscle and the *edl* (*extensor digitorum longus*), a fast-glycolytic muscle. The paper proceeds as follows: Section II describes the experimental design, data are analyzed in Section III, and, finally, Section IV presents the discussion and draws the conclusions.

II. MATERIALS AND METHODS

A. Experimental protocol

30 adults male Sprague-Dawley rats (6 weeks old, average body weight 200 g) were housed in a plastic cage and maintained on a 12L:12D cycle and temperature maintained at 23°C. They were randomly divided into three groups of 10 each: control, CPF1 and CPF5. Rats in CPF1 and CPF5 treated groups were fed a diet containing the CPF at 1 mg/kg/day and 5mg/kg/day respectively for 6 weeks. The control group was given only the standard pellet diet with the vehicle (corn oil).

B. Behavioral experiments

The locomotor activity was assessed on a weekly basis during the treatment period. In the beam walking test, the rat had to cross the beam readily in order to reach a box placed at the other end. Traversing latency and number of hind-limb slips were measured. In the prehensile traction test, rats were assessed for their ability to grasp with their forepaws a horizontal rod, for a one minute. Latency from when the animal mounted the rod to when it fell from it was measured.

C. Measurement of muscle contractility parameters

After 6 weeks of exposure to CPF, the rats were sacrificed. Soleus and *edl* were removed and the preparation was mounted as described by Joumaa and Leoty [6]. The preparation was stimulated by square electrical pulses at 0.1 Hz. After equilibration, single twitches were elicited. For each twitch the peak twitch tension was determined. The fatigue resistance was evaluated using a protocol consisting of a 2 Hz train of supramaximal stimuli for a 5 min period. The fatigue index was calculated as the percentage of initial force remaining after 5 min of muscle stimulation.

III. RESULTS

A. Behavior measurements

Our results show that at weeks 5 and 6, there was a significant increase in time to cross the beam in both CPF1 and CPF5 compared to control group (Fig. 1).

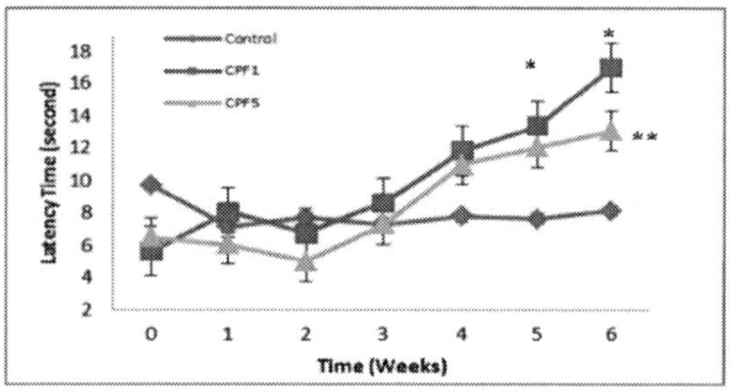

Figure 1. Beam walking test/ latency time. The data are depicted as mean ± SEM. *: $p < 0.05$ for CPF1 vs control at week 05 and 06 and CPF5 vs control at week 05. **: $p < 0.01$ for CPF5 vs control at week 06.

Also, in the hind limb slips test, there was a significant increase in the mean number of errors in both CPF1 and CPF5 groups compared to control group. Regarding the prehensile traction test, there was a significant decrease in latency time from when the animal grasped the horizontal rod with its forepaws to when it fell from it between treated groups (CPF1, CPF5) and control group at weeks 5 and 6 (Fig. 2). No significant difference was observed between CPF1 and CPF5 groups at weeks 5 and 6.

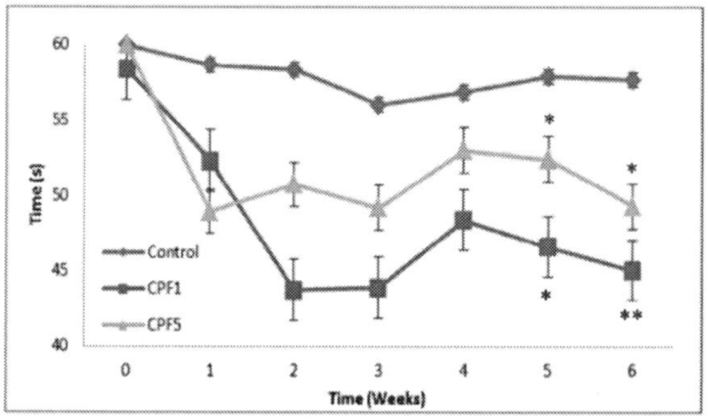

Figure 2. Beam balance test. The data are depicted as mean ± SEM. *: $p < 0.05$ for CPF1 vs control at week 05 and CPF5 vs control at week 05 and 06. **: $p < 0.01$ for CPF1 vs control at week 06

B. Contractility parameters

For the *edl* muscle, our data showed a significant decrease in twitch of contraction (Pt) in CPF1 rats compared to control rats. However, a significant increase in twitch of contraction was observed in rats exposed to CPF5 (Table I). In Soleus muscle, twitch of contraction significantly increase in both CPF1 and CPF5 rats compared to control rats (Table I). Repeated low level exposure to both CPF1 and CPF5 significantly increases the fatigability index of soleus but not *edl* muscle.

TABLE I. EFFECTS OF CPF1 AND CPF5 EXPOSURE ON CONTRACTILE CHARACTERISTICS OF EDL AND SOLEUS MUSCLES. PT PEAK TWITCH TENSION. *: P < 0.05. THE DATA ARE DEPICTED AS MEAN ± SEM.

	Pt (g/cm2)	
	EDL	Soleus
Control (n=10)	150.38 ± 7.96	106.09 ± 4.72
CPF1 (n=10)	96.55 ± 7.32*	133.64 ± 5.08*
CPF5 (n=10)	185.56 ± 9.19*	173.44 ± 6.86*

IV. DISCUSSION AND CONCLUSION

The effect of CPF on locomotor activity has been described previously but mainly during acute exposure and to less extent during chronic repeated low level of exposure. The present study revealed that animals treated with CPF showed deficits during locomotion dose independent following beam walking test, which have been clearly manifested as an increase in latency time, as well as the number of errors through walking the horizontal rod. Impairment in the motor function was also recorded following the prehensile test showing a significant decrease in the time spent and the ability for hanging on the rope compared to the control group. This alteration in locomotors activity can be referred either to an effect on CNS or peripheral skeletal muscles. In our study, the dose independent decrements reported in the behavioral tests of treated rats came along a significant effect on the soleus and edl muscle's amplitude of contraction and fatigability. The greater level of contraction and the decrease in fatigue resistance in vitro for soleus muscle at the two different doses might be linked to the alterations shown in the locomotors activity of rats. Our present findings are consistent with the documented effects of repeated low level exposure to CPF on another slow oxidative skeletal muscle which is the diaphragm [7]. The increase in twitch tension is explained by the decrease of AchE activity in the studied muscle [7]. In conclusion, the significant variations in motor activity and contractile parameters suggest a possible effect of CPF on skeletal muscle function by affecting the expression of some of the main actors of contraction, such as Ryanodine receptor and SERCA pump that are responsible for calcium homeostasis. This study demonstrates that exposure to CPF has an impact on the vital physiological functions of vulnerable populations. These deleterious effects could have long-term consequences in maintaining body homeostasis and human health.

REFERENCES

[1] Grand view research, " Chlorpyrifos Market Analysis, Market Size, Application Analysis, Regional Outlook, Competitive Strategies, and Forecasts, 2015 To 2022". 2014, http://www.grandviewresearch.com/industry-analysis/chlorpyrifos-market [retrieved: September, 2018]

[2] C. Timchalk et al. "A Physiologically based pharmacokinetic and pharmacodynamic (PBPK/PD) model for the organophosphate insecticide chlorpyrifos in rats and humans," Toxicol. Sci., vol. 66, pp. 34–53, 2002.

[3] T. K. Chakraborti, J. D. Farrar, and C. N. Pope, "Comparative neurochemical and neurobehavioral effects of repeated chlorpyrifos exposures in young and adult rats,". Pharmacology Biochemistry and Behavior, vol. 46, pp. 219-224, 1993.

[4] W. Chen, Y. Zhang, L. Yuan, Y. Li, and J. Li, "Neurobehavioral evaluation of adolescent male rats following repeated exposure to chlorpyrifos," Neuroscience Letters, vol. 570, pp. 76–80, 2014, doi/10.1016/j.neulet.2014..03.068.

[5] C. Yan, L. Jiao, J. Zhao, H. Yang, and S. Peng, "repeated exposures to chlorpyrifos lead to spatial memory retrieval impairment and motor activity alteration," Neurotoxicology and teratology, vol. 34, pp. 442-449, 2012, doi:10.1016/j.ntt.2012.05.053

[6] W. H. Joumaa and C. Leoty, "Differential effects of nandrolonedecanoate in fast and slow rat skeletal muscles," Med Sci Sports Exerc, vol 33, pp. 397-403, 2001.

[7] W. Darwiche et al. "Impact of chronic exposure to the pesticide chlorpyrifos on respiratory parameters and sleep apnea in juvenile and adult rats," PLoS ONE, vol. 13, pp. 1-14, 2018, doi./10.1371/ journal.pone.0191237.

Impact of Resistance Training in Patients with Chronic Heart Failure

Zahra Sadek/ Said Ahmaidi
Laboratory EA 3300 (APERE) « Adaptations Physiologiques à l'Exercice et Réadaptation à l'Effort »
Picardie de Jule Vernes University
Amiens, France
Zahra.sadek@etud.u-picardie.fr
Said.ahmaidi@u-picardie.fr

Wissam Joumaa / Charifa Awada
Laboratory « PhyToxE Physio-toxicité environnemental »
Faculty of Sciences, Lebanese University
Nabatieh, Lebanon
wjoumaa@ul.edu.lb
charifa.awada@hotmail.com

Mahmoud Youness
Medical Research Center
Beirut Cardiac Institute
Beirut, Lebanon
Myouness55@yahoo.com

Wiam Ramadan
Lebanese Institutes for Biomedical Research and Application (LIBRA)
International University of Beirut (BIU)
Beirut, Lebanon
Wiam.ramadan@liu.edu.lb

***Abstract*— This study aims to determine the positive effects of Resistance Training in Chronic Heart Failure patients with inspiratory muscle weakness. 16 patients were enrolled, and underwent a 3 months training program. They were randomized into one of two groups consisting of Resistance Training or control group. Resistance Training was performed at 60% of 1 repetition maximum and each workout was composed of 3 sets of 10 repetitions, with at least 30 seconds of rest between sets. However, the control group did not participate in any exercise program. Pulmonary Function, Respiratory and Skeletal Muscles Function tests were performed as well as Quality of Life and Functional Capacity before and after training periods. Results have shown the beneficial effects of Resistance Training on the skeletal and respiratory muscle function, the functional capacity reflected by the six-minute walk test, the cardiac function, exercise performance, as well as on dyspnea and Quality of Life.**

Keywords- Heart Failure; Resistance Training; Skeletal Muscle Function; Ejection Fraction.

I. INTRODUCTION

Exercise intolerance, dyspnea and fatigue are the main obstacles that Chronic Heart Failure (CHF) patients face during their daily life activities. These phenomena might contribute to physical impairment and result in a poor Quality of Life (QoL) [1]. Such restrictions happen as a consequence of reduction in skeletal muscle mass and strength, which might be explained by both qualitative and quantitative abnormalities [2]. In fact, there are histological and biochemical derangements expressed respectively by altered fibers distribution and reduced oxidative enzyme activity in addition to muscle metabolism impairment, mitochondrial changes, inflammation and muscle atrophy [3]. After a period of intense evaluation of the safety and effectiveness of exercise rehabilitation in CHF patients, exercise training was shown to be the cornerstone of cardiac rehabilitation programs.

Because muscle dysfunction represents a hallmark of heart failure, the emphasis was on Resistance Training (RT), in order to restore the normal muscle structure and function [4]. For many years, bed rest and limited physical activity were recommended for all stages and forms of heart failure; while exercise was not suggested [5]. Nowadays, however, the concept of cardiac rehabilitation, including exercise training, and specifically RT, is well spotlighted and highly recommended recently, mainly because of studies that show its benefits in various outcomes [6] [7]. In fact, application of such programs induces significant histological, metabolic and functional adaptations in skeletal muscles, thereby, improving patient's QoL. McKelvie was the first to demonstrate in 1995 that there are no significant differences between cycling and RT regarding left ventricle response in heart failure patients [8]. This study found similar results with those obtained from Meyer et al. where central hemodynamics was stable and well tolerated during resistance exercise [9]. In addition, from the research conducted by Grosse et al. who performed RT at 65% of 1 Repetition Maximum (RM), an increase of 80–102% of muscular endurance and 14.5% of VO2peak was reported [10]. Pu et al. who performed RT at 80% of 1RM, also showed a 43% and 13% increase of muscular strength and Six-Minute Walk Test (6MWT), respectively [11].

These findings are consistent with those of Levinger et al. who discovered 18% increase of muscular strength and 19% amelioration of VO2 peak after training patients at 40–80% of 1RM [12]. As shown above, the significant improvements in muscle strength and endurance, the adaptation of muscle mass and the increase in the QoL and functional capacity had been proved by many researchers after RT [11] [13] [14]. Each study performs RT according to specific characteristics such as intensity, duration, frequency, number of repetitions and sets of exercise. Overall, these features should be taken into consideration to avoid any cardiovascular stress, and thereby any harmful consequence. The aim of our study is to determine the effects of RT on skeletal and respiratory muscle function, functional capacity, cardiac function, dyspnea and QoL in patients with CHF. The paper proceeds as follows: Section II describes the experimental design, data are analyzed in Section III, Section IV presents the discussion and, finally, Section V draws the conclusions.

II. Methods

16 patients with stable CHF and inspiratory muscle weakness were randomly assigned to a training program for 12 weeks (3 times / week). The patients were divided, thereafter, into two different groups: controls (n=8) and resistance (n=8). Patients in the control group did not exercise at all and were instructed to continue their normal life activities during the three months of trial. However, RT consisted of strength exercises that targeted the muscles of the quadriceps, the hamstrings and gluteus muscles of the lower extremities; Biceps, Triceps, Deltoid, and Pectoralis of the upper extremities. The patients started training at 60% of 1RM that was assessed and recalculated every two weeks.

During the test, verbal encouragement was used. Maximum Voluntary Isometric Force (MVIF) for right and left muscles was measured; three sets of three repetitions, separated by 20 seconds of rest were given to the patient to develop maximum force.

Concerning the isometric endurance time or Maintenance Time (MT), it was assessed by maintaining 50% of the MVIF as long as possible, until exhaustion. This procedure was repeated three times separated by 5 minutes of rest.

At baseline and after the training period, patients underwent pulmonary function test by spirometry, respiratory muscle function assessment by electronic pressure transducer in order to measure Respiratory Muscle Strength (MIP), echocardiography, to asses Left Ventricle Ejection Fraction (LVEF), stress test, skeletal muscle function test using hand-held dynamometer, and 6MWT. Dyspnea, according to Borg scale, and QoL, according to Minesotta living with heart failure questionnaire (MLWHFQ), were also assessed.

III. Results

Compared to control, the training group had 18% ($p<0.05$) and 11% ($p<0.05$) improvement in the right and left MVIF respectively, and a 24% improvement in the right quadriceps muscle endurance capacity (MT) ($p <0.05$).

Moreover, RT has shown a 27% improvement in exercise time ($p<0.01$), 24% in METS ($p<0.01$), and a 15% improvement in dyspnea sensation ($p<0.001$). A significant increase in 6MWT distance (13%, $p<0.01$), a decrease in NYHA functional class (33%, $p<0.05$), and a decrease in MLWHFQ score (30%, $p<0.05$) were also noticed. In addition, a little increase in (LVEF) was observed in RT (5%, $p<0.05$) versus no changes in the control group. Concerning respiratory muscle function, RT was able to improve MIP by 14% ($p<0.01$).

IV. Discussion

In this study, RT was able to bring out significant improvements in the skeletal muscle function, the cardiac function, NYHA functional class, and dyspnea as well as on functional capacity and QoL. Surprisingly, RT has also shown significant impact on respiratory muscle strength. Since the control group had no significant improvements in any of these parameters, we can confirm the effectiveness of the RT.

The benefits gained in skeletal muscle function affect positively the overall exercise performance and are closely related to beneficial adaptations in the muscle structure and function such as an increase in type 1 fiber, decrease in circulatory pro-inflammatory markers and a better muscle oxidative capacity [4]. In addition, such an intensive RT might induce an increase in the motor unit recruitment and so will impact on root mean square (RMS) value towards an upward trend.

In addition, we observed that the right quadriceps muscle is stronger than the left quadriceps muscle. These variations can be explained by the fact that maybe most of patients are right leg dominants.

Besides, as known, skeletal and respiratory muscle changes in heart failure are associated with biochemical and metabolic disorders. Thus, the fact of improvement in respiratory muscle function in our study may be attributed to the beneficial effects of RT in increasing mitochondrial enzymes and decreasing pro-inflammatory cytokines [3]. The skeletal muscle hypothesis confirms that impairments in the skeletal muscle not only alter the skeletal muscle by itself, but also contribute to further deteriorations and worsens the symptomes [15]. In the same manner as this hypothesis, RT also works not only by improving the skeletal muscle function, but also by enhancing the overall exercise performance and QoL, thus reducing hospitalizations as well as mortality rates. Therefore, regular RT programs are very efficient in counteracting these negative skeletal muscle abnormalities seen in CHF.

V. Conclusion

This study had shown the crucial effect of RT, for it has improved the respiratory muscle strength. RT was safe and effective in improving skeletal muscle function, exercise performance, dyspnea and QoL in CHF patients, as well as in improving cardiac LVEF.

We recommend that future studies investigate the usefulness of the electrical activity of the muscles known as Electromyography (EMG) in clinical diagnosis, in heart failure patients, in order to monitor the progression of skeletal muscle activity and function.

ACKNOWLEDGMENT

This research was supported by a fund from the National Council for Scientific Research (CNRS), Beirut, Lebanon; and was approved by the ethics committee of Beirut Cardiac Institute, Beirut, Lebanon.

REFERENCES

[1] M. Grazzini, L. Stendardi, F. Gigliotti, and G. Scano, "Pathophysiology of exercise dyspnea in healthy subjects and in patients with chronic obstructive pulmonary disease (COPD)," Respir. Med., vol. 99, no. 11, pp. 1403–1412, 2005.

[2] M. F. Piepoli and A. J. S. Coats, "The 'skeletal muscle hypothesis in heart failure' revised," Eur. Heart J., vol. 34, no. 7, pp. 486–488, 2013.

[3] C. Zizola and Schulze P. Christian, "Metabolic and structural impairment of skeletal muscle in heart failure," vol. 6, no. 5, pp. 247–253, 2009.

[4] C. Delagardelle, P. Feiereisen, P. Autier, R. Shita, R. Krecke, and J. Beissel, "Strength/endurance training versus endurance training in congestive heart failure," Med Sci Sport. Exerc, vol. 34, no. 12, pp. 1868–1872, 2002.

[5] C. D. McDonald, G. E. Burch, and J. J. Walsh, "Prolonged Bed Rest in the Treatment of the Dilated Heart," Circulation, vol. 32, no. 5, pp. 852–856, 1965.

[6] P. Ponikowski et al., "2016 ESC Guidelines for the diagnosis and treatment of acute and chronic heart failure," Eur. Heart J., vol. 37, no. 27, p. 2129–2200m, 2016.

[7] C. W. Yancy et al., "2013 ACCF/AHA guideline for the management of heart failure: A report of the American college of cardiology foundation/american heart association task force on practice guidelines," J. Am. Coll. Cardiol., vol. 62, no. 16, 2013.

[8] S. M. Robert, L. Nei, T. Charles, B. Robert, and M. J.Duncan, "Comparison of hemodynamic responses to cycling and resistance exercise in congestive heart failure secondary to ischemic cardiomyopathy," Am. J. Cardiol., vol. 76, no. 12, pp. 977–979, 1995.

[9] K. Meyer et al., "Hemodynamic responses during leg press exercise in patients with chronic congestive heart failure.," Am. J. Cardiol., vol. 83, no. 11, pp. 1537–43, 1999.

[10] K Kreulich, and H Nagele, "Peripheral mus-cular strength training in patients with severe heart failure," Dtsch Z Sport., vol. 52, pp. 11–14, 2001.

[11] C. T. Pu et al., "Randomized trial of progressive resistance training to counteract the myopathy of chronic heart failure," J Appl Physiol, vol. 90, no. 6, pp. 2341–2350, 2001.

[12] I. Levinger, R. Bronks, D. V. Cody, I. Linton, and A. Davie, "Resistance training for chronic heart failure patients on beta blocker medications," Int. J. Cardiol., vol. 102, no. 3, pp. 493–499, 2005.

[13] I. Levinger, R. Bronks, D. V Cody, I. Linton, and A. Davie, "Perceived exertion as an exercise intensity indicator in chronic heart failure patients on Beta-blockers.," J. Sports Sci. Med., vol. 3, no. YISI 1, pp. 23–7, 2004.

[14] G. Magnusson et al., "High intensity knee extensor training, in patients with chronic heart failure. Major skeletal muscle improvement," Eur Hear. J, vol. 17, no. 7, pp. 1048–1055, 1996.

[15] A. L. Clark, P. A. Poole-Wilson, and A. J. S. Coats, "Exercise limitation in chronic heart failure: Central role of the periphery," J. Am. Coll. Cardiol., vol. 28, no. 5, pp. 1092–1102, 1996.

Eco-Sensors4Health Toolkit: Scaffolding Children Participation in Schools' Environmental Health

Maria João Silva
School of Education
Polytechnic Institute of Lisbon, Portugal
Lisbon, Portugal
e-mail: mjsilva@eselx.ipl.pt

Eduarda Ferreira
CICS.NOVA
New University of Lisbon
Lisbon, Portugal
e-mail: e.ferreira@fcsh.unl.pt

Alexandra Souza, Ana Rita Alves
Ciência Viva School
Pavilion of Knowledge
Lisbon, Portugal
e-mail: {asouza, aalves}@cienciaviva.pt

***Abstract*— The research presented in this paper aims at improving health education in primary schools, supporting the use of environmental sensors in environmental health problem solving. A research based design approach was adopted to develop a toolkit, aiming at supporting teachers in scaffolding children's participation to improve their schools' environmental health. The toolkit will facilitate children's activities that make use of eco-sensors to identify and explore environmental health problems, as well as to suggest local solutions. Sound pollution, air pollution, and lack of thermal comfort are three central problems in Portuguese primary schools' environmental health. The data acquired by children through the use of sensors will inform and validate children's decisions to improve schools' environmental health and will be made available to other schools in a collaborative platform. The design of the toolkit was situated in Portuguese primary schools. It is an iterative design process, based on teaching and learning interventions with primary school children, in-service teachers, and students of a teacher education graduation, all in their natural and complex settings. Those interventions confirmed the practicality and effectiveness of the planned activities, and, together with the results of previous projects, informed the content and form of the designed toolkit that will be used in diverse schools to improve schools' environmental health.**

Keywords – environmental health; eco-sensors; children; participation.

I. INTRODUCTION

A. *Thematic, research goals, and context*

Primary schools' environmental health is a current and relevant subject in Portugal, since there are several identified problems that affect school communities' well-being, and that consequently can affect the learning processes [1] – [3]. Sound pollution, air pollution, and lack of thermal comfort are three central problems in Portuguese primary schools' environmental health [1] – [3]. These problems are three important dimensions of indoor environmental quality [2] that, together or separately, create concentration difficulties [1], and consequently affect students' and teachers' performance [2].

Eco-Sensors4Health is a research project whose main goal is to improve environmental health in Portuguese primary schools, through the use of eco-sensors by children. Eco-sensors are devices that detect and measure environmental factors and elements, such as temperature, sound, and carbon dioxide. Eco-sensors communicate those measurements to other electronic devices, which transform the sensors' signals into multiple representations, such as digits, tables, and graphs.

In the Eco-Sensors4Health project, children's participation to improve schools' environmental health includes: i) the use of eco-sensors to identify and explore problems; ii) the suggestion of local solutions that will also be validated with the use of eco-sensors. This way, children's participation in the Eco-Sensors4Health project will be developed in the context of primary school curricular activities, using techniques of the scientific inquiry strategy.

The data acquired by children with the eco-sensors will be made available on the Eco-Sensors4Health collaborative platform, allowing multiple queries and comparisons of environmental health conditions in different schools and circumstances.

The research presented in this paper is part of the Eco-Sensors4Health project. It aims at improving health education in primary schools, through the creation of a toolkit, using a research based design approach, to support primary school teachers in scaffolding children's participation in schools' environmental health.

B. *Goals of the Eco-Sensors4Health toolkit*

The Eco-Sensors4Health toolkit will support monitoring and intervention in environmental health, enabling children to eco-innovate to create healthier environments in schools. Since it is widely recognized that mediation, namely in what concerns scaffolding, is fundamental in problem solving and scientific inquiry activities [4] [5], this toolkit is targeted to primary school teachers, to facilitate the implementation of children's activities. It will offer support to teachers in scaffolding children in the use of eco-sensors to identify and understand school environmental health problems, allowing the suggestion of solutions to such problems.

The specific goals of the Eco-Sensors4Health toolkit are:

- To make eco-sensors and tablets available to teachers and children in order to support the acquisition of schools' environmental health data;
- To offer suggestions and support documents for problem solving and scientific inquiry activities. These suggestions and documents should also make it easier for teachers and children to enter data in the Eco-Sensors4Health platform;
- To make pre and posttests available to teachers (questionnaires to support learning assessment).

Following this introduction (Section 1), the theoretical background, and the related work are presented in Section 2. Then, in Section 3, the methodology, and toolkit design process are described. The subsequent sections are the systematization of the lessons learned and of the conclusions (Section 4).

II. THEORETICAL BACKGROUND AND RELATED WORK

The design and development of the Eco-Sensors4Health toolkit represent a development study, in the context of an educational design research, since it aims at developing research-based solutions for complex problems in educational practice [6]. This section reports the main dimensions of the first phase of this development study: the review of literature.

The relevance of monitoring and intervention in schools' environmental health is highlighted by diverse authors, emphasizing that:

- The school's environment can affect the health of the school community [7], with children being particularly vulnerable to environmental health risks, such as air pollution [8];

- Environmental health in elementary schools is often inadequate, especially when financial resources are scarce, and the ventilation is insufficient with classroom temperatures reaching values out of the recommended range, and carbon dioxide levels exceeding the safety level [9] [10].

The design and development of the Eco-Sensors4Health toolkit are grounded in the recognition of the importance of the participation of children in schools' environmental health, since the participation of students in health promotion has demonstrated positive personal effects on students, as well as positive effects on the school [11].

The participation of children in schools' environmental health, using problem solving and scientific inquiry strategies, needs scaffolding, since children need support (guiding and resources [12]) to succeed in such activities [5]. In this context, scaffolding should integrate concreteness fading over time, and transfer of responsibility to students [13]. In concreteness fading, the concreteness of the representations successively decreases to allow students to understand abstract representations that are connected to the concrete situation that is represented [14]. Following this scaffolding guiding principles, in the Eco-Sensors4Health toolkit, activities are sequenced from sensory exploratory activities to data acquisition in inquiry and problem solving activities. Since observation tables and experiment plans can make scientific strategies visible and transfer responsibility to students, the Eco-Sensors4Health toolkit will include such tables and plans, together with sensors, as manipulable mediators [5] [12] – [15].

Electronic sensors, together with data-loggers, can be cognitive tools in scaffolding, facilitating the complex task of monitoring environmental variables, and improving the level of children's scientific reasoning [4]. Diverse related projects evidenced the practicality of the use of sensors by elementary school children in environmental and health sense making activities [16]. In the SchoolSenses@Internet project, children used mobile phones built-in sensors, to create multiple sensory views of environmental quality, portraying and assessing the schoolyard environment [17]. In the USense2Learn project, children used mobile phones built-in sensors together with external weather sensors to the mobile creation of georeferenced multisensory information, intertwining quantitative and qualitative visions of the schoolyard, and sharing such information with other classrooms, using Google Earth [18]. Also, with a context aware approach, the SENSE project engaged elementary school students in the use of environmental sensors to collect, analyse, reflect, and share authentic air pollution data [19].

In order to support elementary school teachers and children, the TEEMSS2 [20] and the POLLEN [21] projects created documentation that describes activities in which children use sensors to explore the environment and its multiple representations.

The scaffolding of the use of sensors by children, in the context of scientific inquiry and problem solving activities, should address the following requisites: i) an appropriate balance of automated technology and user control [22]; ii) contextualization of data acquisition [19]. Consequently, the Eco-Sensors4Health toolkit adopts: i) a manual approach to data acquisition (children will register sensors' data manually in registration forms and in the platform), following the affordances and challenges identified in previous studies and projects [18] [21]; ii) the manual registration of time and local data in the registration forms; iii) and the production of photos/audios/videos and captions to contextualize the acquired data [19].

III. THE TOOLKIT DESIGN PROCESS

The design process of the Eco-Sensors4Health toolkit is iterative, based on teaching and learning interventions, situated and implemented in real world contexts (Portuguese primary schools). The research team includes primary school teachers, together with experts in education, in environmental health, and in information and communication technologies.

The first version of the Eco-Sensors4Health toolkit is based on the conceptual framework, drafted in the previous section. To acquire empirical evidences of the practicality and effectiveness of this version, three case studies were developed with different participants: primary school children, pre-service teachers, and in-service teachers.

A. *Structure of the first version of the toolkit*

Following the three central problems of the Portuguese schools' environmental health [1], the specific goals presented in the introduction, and the conceptual framework drafted in the previous section, the main components of the Eco-Sensors4Health toolkit are:

- A set of tablets, iOS or Android, with an app to allow the sensors data logging;
- A set of sensors: Sound sensor (tablets have in-built sound sensors); Carbon dioxide sensor (PASPORT Carbon Dioxide Gas Sensor - PS-2110); Temperature and humidity sensor (PASPORT Weather Anemometer Sensor - PS-2174);
- Guidelines for children's activities on schools' environmental health, with the following structure: i) introduction to main concepts; ii) exploring with the senses (research questions; tasks; questions for reflection); iii) measuring and interpreting (research questions; tasks; questions for reflection); iv) suggesting solutions;
- The registration forms and collaborative documents (experiment plans) to support children's activities on schools' environmental health;
- The children's pre and posttests (with questions in the following categories: Knowledge, Environmental and Health Awareness, Attitudes, Personal Investment and Responsibility, Perception of the Physical Environment).

B. *Teaching and learning interventions with primary school children*

The empirical testing was developed with primary school classes in the Ciência Viva School (CVS), a science museum school in Lisbon. Each class stays in CVS for a week to develop a technology enhanced experiential science learning program. This program is mediated by the CVS teachers/researchers, whereas the school teachers of the classes follow all the process. CVS was the context chosen by the Eco-Sensors4Health project to develop empirical testing, since it made it possible to develop teaching interventions with two visiting classes in each week (60 classes each year).

Two research questions were formulated: i) Do children perform epistemic practices in a scientific inquiry that makes use of the Eco-Sensors4Health toolkit? ii) Can the Eco-Sensors4Health toolkit be used in a scientific inquiry to support children in identifying school's environmental health problems and suggest solutions?

Using the toolkit, two different teaching and learning interventions were successfully implemented during the school year of 2017-2018: the "Sound Pollution", and the "Air Pollution" interventions. The implementation of the "Sound Pollution" intervention in two classes was observed and audio-recorded, while the "Air Pollution" intervention was observed and audio-recorded with another class.

Epistemic practices (EP) are practices that "construct scientific knowledge, having as reference the scientific practices in the context of scientific production" [23]. The audio-recordings allowed the identification of diverse types of EP, by project researchers with experience in such identification. Some examples of the EP, performed by children during the measurement and interpretation tasks, are presented in Table I. These EP were facilitated by teacher mediation (performed by teachers/researchers) [23], and guided by research questions, such as: "How does carbon dioxide concentration change when we change our location in the school environment?", "How does carbon dioxide concentration change when we open the door/window of the classroom?", "How does sound level change when we change classroom activities?", or "How does sound level change when we change our location in the school environment?" Likewise, it was possible to identify these same types of EP during the sensory exploration of sound and air tasks, which were also guided by teacher mediation, and research questions (e.g., "Can we observe the sound waves?" or "Can we catch the air?").

TABLE I. EXAMPLES OF EPISTEMIC PRACTICES

Epistemic practices	Examples
Describe	After opening the door of the classroom, children read the values of carbon dioxide concentrations displayed by the sensor and reported the decreasing in time (1650 ppm, 1480 ppm, 1223 ppm, 913 ppm, 580 ppm...)
Forecast	When asked if the values that the sound sensor will register will be higher while singing the Happy Birthday or while clapping, children gave different answers, justifying their estimations.
Use sensors	Children used the carbon dioxide sensor near the road, and measured at first a concentration of 450 ppm, and afterwards a concentration of 5000 ppm, after a car passed by.
Interpret	In the sequence of the carbon dioxide data acquisition near the road, a child verbalized that when a car passes by, the values increase.
Organize information	Children registered the data acquired with the sensor in the registration form.
Relate	When asked why were the measured values of carbon dioxide concentrations lower in the garden, a child answered that it was because there are more plants in the garden.

In both "Sound Pollution" and "Air Pollution" interventions, all children were able to fill the registration forms with the data acquired with sensors. Figure 1 shows the "Air Pollution" registration form, filled by a child. It is possible to observe the carbon dioxide concentrations in two days (columns), in the empty classroom, during the class, after opening the classroom door, and in the exhaled air (lines). It is also possible to observe the concentrations outdoors, at the Pavilion door, in the garden and near the road. In this context, after gathering the sensors' data in different conditions, the great majority of children were able to make sense of the data, suggesting ways of renewing the air in the classroom: opening the door or the window (last sentence completed on the form). In what concerns "Sound Pollution" intervention, children were also able to gather and register sensors' data in different conditions, during different activities. In one class of the "Sound Pollution" intervention, children suggested ways to improve hearing protection, but in the other class, children only repeated the sentence suggested by the teacher/researcher. This result will inform teachers/researchers to avoid this kind of influence.

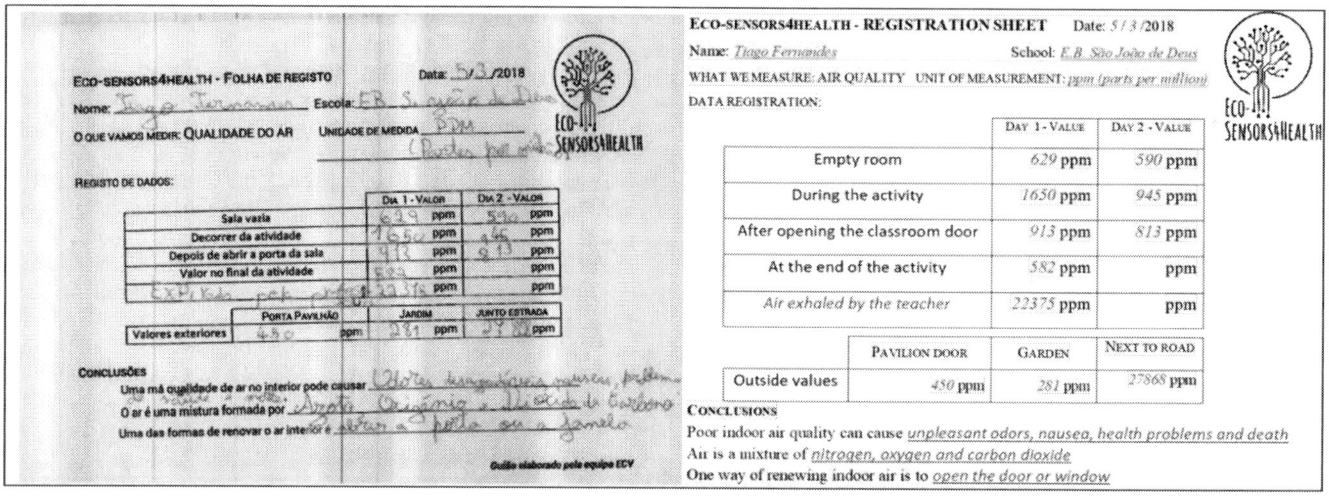
ECO-SENSORS4HEALTH - REGISTRATION SHEET Date: 5 / 3 /2018

Name: Tiago Fernandes School: E.B. São João de Deus

WHAT WE MEASURE: AIR QUALITY UNIT OF MEASUREMENT: ppm (parts per million)

DATA REGISTRATION:

	Day 1 - Value	Day 2 - Value
Empty room	629 ppm	590 ppm
During the activity	1650 ppm	945 ppm
After opening the classroom door	913 ppm	813 ppm
At the end of the activity	582 ppm	ppm
Air exhaled by the teacher	22375 ppm	ppm

	Pavilion door	Garden	Next to road
Outside values	450 ppm	281 ppm	27868 ppm

CONCLUSIONS

Poor indoor air quality can cause unpleasant odors, nausea, health problems and death

Air is a mixture of nitrogen, oxygen and carbon dioxide

One way of renewing indoor air is to open the door or window

Figure 1. Close view of an "Air Pollution" registration form filled out by a child (on the left, the original filled registration form; on the right, the translation into English)

In spite of the efforts of the teachers/researchers to avoid the influence of the school teachers of the classes in children's answers, these classes' teachers helped some of the children in answering some of the questions, especially in the pretest. This influence was noticed but could not be quantified.

It is noteworthy that in both interventions, and in both pre and posttests, children's answers to 'Environmental and Health Awareness' questions were very positive. Children's answers to 'Knowledge' questions related to the topic of each intervention improved after the respective intervention (see Table II). For instance, in the "Sound pollution" intervention, children's answers to the question "A strong sound level is harmful to health?" had a significant difference (improvement) between the pre and posttest (Wilcoxon= - 2.639, $p<.05$). However, in the "Air pollution" intervention, children's answers to the same question did not have a significant difference (Wilcoxon= -0.312, $p>.05$). Furthermore, in the "Air pollution" intervention, children's answers to the question "To ventilate the classroom is important for health?" had a significant difference (improvement) between the pre and posttest (Wilcoxon= - 2.627, $p<.05$). However, in the "Sound pollution" intervention, children's answers to the same question did not have a significant difference (Wilcoxon= -0.550, $p>.05$).

TABLE II. STATISTIC DATA OF 'KNOWLEDGE' QUESTIONS RELATED TO THE TOPIC OF EACH INTERVENTION

"Sound pollution" intervention			
Question	Average		Wilcoxon
	Pre	Post	
"A strong sound level is harmful to health?"	3.34	4.29	-2.639 p<.05
"Air pollution" intervention			
Question	Average		Wilcoxon
	Pre	Post	
"To ventilate the classroom is important for health?"	4.15	4.67	-2.627 p<.05

Unexpectedly, the children of the classes that participated in the "Sound Pollution" intervention did not improve the answers to the question "When we are silent, we do not hear sounds in the classroom?", from pre to posttests. It will be relevant to verify if those results are due to the negative formulation of the sentence, since children answered correctly to equivalent questions in debates during the intervention.

Several teachers of the participating classes continued to support children in monitoring and solving noise problems in their own school. Some teachers asked for support in implementing that intervention to the whole school.

C. *Teaching and learning interventions with students of a teacher education graduation*

To assess the perceived utility of sensors to study schools' environmental health problems, teaching and learning interventions were performed in a Statistics course and in a Biology course of a teacher education graduation school. The selection of these courses was based on two criteria: i) adequate course syllabus; ii) large number of participating students.

The intervention in the Statistics course aimed at understanding the perceived pertinence of the use of the sensors of mobile phones to acquire sound level data in diverse locations of the school, and of the use of Excel to organize and process the data. The 92 (100%) participants showed no technical difficulties in using the app of the mobile phones to acquire sound level data, nor in making sense of the acquired data. 82% of the participants found the activity pertinent, mainly (36%) because of the topic (sound), and also because of the opportunity to consolidate statistics knowledge (20%).

The intervention in the Biology course was developed with the same participants working in groups and aimed at understanding the perceived pertinence of the use of the carbon dioxide sensor to study the exchanges of that gas between the atmosphere and living beings.

The content analysis of the groups' reports allowed the formative assessment of the perceived pertinence of the intervention. 84% of the groups reported that the intervention contributed to their knowledge of scientific concepts and phenomena, while 34% mentioned that the intervention was pertinent because it contributed to the development of didactic competences (use of the sensors with their future students). 84% of the groups wrote about the pertinence of the use of sensors with children, emphasizing the enhancement of the learning process, namely in what concerns knowledge acquisition (69%), and the increase of motivation (28%). Nonetheless, the analysis of the reports showed students' difficulties in learning complex phenomena, such as carbon dioxide exchanges, respiration and photosynthesis.

D. *Teaching and learning interventions with in-service teachers*

A b-learning workshop with in-service teachers was developed by the research team to assess the utility and practicality of the Eco-Sensors4Health toolkit. The following research questions were formulated: Can teachers use the Eco-Sensors4Health toolkit to develop teaching and learning interventions, in which children: i) identify school's environmental health problems? ii) Suggest solutions to the identified problems?

Nine teachers participated in the workshop. Five teachers implemented the interventions individually and four worked in two groups. However, only six interventions were presented by eight teachers: four interventions were centered on temperature, while two were centered on noise.

Teachers showed evidences that in all the implemented interventions, children (from 2nd to 5th grade): i) started by the sensory exploration of temperature/sound; ii) used sensors to measure temperature/sound in the schools' environment; iii) identified noise and thermal discomfort problems in specific locations and activities; iv) suggest solutions to the identified problems.

Using the sensors and temperature/sound scales provided in the toolkit, the children recognized harmful sound levels (e.g., 83 dB in the canteen at lunch time) and non-comfort temperatures (e.g., 26.4°C in the classroom). They were able to suggest solutions, such as "keep the classroom open and ventilated", "use light shutters", "use the sound sensor to be aware of the sound level", and "play 'silence games' in the canteen".

All the teachers expressed the will to continue the interventions in their classes.

IV. LESSONS LEARNED AND CONCLUSIONS

The Eco-Sensors4Health toolkit aims at supporting teachers in scaffolding elementary school children participation in schools' environmental health. It was developed using a research based design approach and, specifically, a development study [6]. The first version of the toolkit was based on the review of the literature, and included: tablets, sensors, guidelines for environmental health scientific inquiry activities, registration forms and experiment plans, pre and posttests. The empirical testing [6] was carried out in three case studies that were focused on the formative assessment of teaching and learning interventions, guided by the toolkit first version.

Both children and teacher participants of all the three cases showed no technical difficulties in using the sensors and tablets to perform the toolkit activities, making sense of the acquired data, while performing the tasks. In all the three cases, all the participants implemented the toolkit activities in an engaged way, and showed motivation to future similar activities, namely in what concerns monitoring and solving environmental problems in their schools.

In the case study developed with primary school children, the practicality and effectiveness of the toolkit were illustrated by children's epistemic practices, by suggestions to solve the identified problems, and by the answers to the tests. Some results were highlighted, in line with previous research: i) the importance of the research questions included in the toolkit guidelines in eliciting epistemic practices [13]; ii) the effectiveness of the registration forms in scaffolding the sense making in the use of sensors [15]; iii) the role of the reflection questions of the toolkit guidelines in scaffolding children's solutions to the identified problems; iv) the need to improve some questions of the pre and posttests, and to prevent the influence of classes' teachers in children's answers to such tests.

In the second case study, the students of the teacher education school acknowledged the pertinence of the activities (suggested in the toolkit), not only to their own content learning, but also as strategies to use with their future students.

In the third case study, the in-service teachers used the guidelines and tools of the toolkit to successfully scaffold primary school children in identifying, exploring, and suggesting solutions to some schools' environmental health problems.

Following the validation of the guidelines and tools of the toolkit first version, future work will include a refinement of the Eco-Sensors4Health toolkit, based on the lessons learned in the empirical testing, and its use, together with the collaborative platform, to improve health education in diverse elementary schools. The final version of the Eco-Sensors4Health toolkit will be made available on the project's website and will be used to support teacher mediation in environmental health scientific inquiry processes, which will allow children to identify, explore and solve problems in their schools.

ACKNOWLEDGMENT

The present paper is part of the Eco-Sensors4Health project (LISBOA-01-0145-FEDER-023235) that is supported by FEDER (PORTUGAL2020) and Portugal State Budget.

REFERENCES

[1] G. von Amann (Coord.), School health program 2015. [Programa de saúde escolar 2015]. Lisboa: DGS, 2015.

[2] L. D. Pereira, D. Raimondo, S. P. Corgnati, and M. Gameiro da Silva, "Assessment of indoor air quality and thermal

comfort in Portuguese secondary classrooms: methodology and results," Building and Environment, pp. 69-80, 2014.
[3] J. Madureira, et al., "Children's health and indoor air quality in primary schools and homes in Portugal: Study design," Journal of Toxicology and Environmental Health, Part A, vol. 78, pp. 915–930, 2015.
[4] B. J. Reiser, "Scaffolding complex learning: the mechanisms of structuring and problematizing student work," Journal of the Learning Sciences, vol. 13, n. 3, pp. 273-304, 2004.
[5] C. Quintana, et al., "A scaffolding design framework for software to support science inquiry," The Journal of the Learning Sciences, vol. 13, n. 3, pp. 337-386, 2004.
[6] T. Plomp, "Educational design research: an introduction," in. Educational design research - part a: an introduction, Plomp, T. and Nieveen, N. , Eds. Enschede, the Netherlands: Netherlands Institute for Curriculum Development (SLO), pp. 10 – 51, 2013.
[7] J. Wargo, The physical school environment: an essential component of a health-promoting school, Geneva, Switzerland: World Health Organization, pp. 1–51, 2004.
[8] World Health Organization (WHO) Children's environmental health, 2018. Available at http://www.who.int/ceh/en/ [retrieved July, 2018].
[9] P. Wargocki and N. A. Da Silva, "Use of visual CO2 feedback as a retrofit solution for improving classroom air quality," Indoor Air, vol. 25, n. 1, 105-14, 2015.
[10] A. M. C. Ferreira, Schools' indoor air quality and children's health. [Qualidade do ar interior em escolas e saúde das crianças]. PhD Dissertation. Coimbra: Universidade de Coimbra, 2014.
[11] U. Griebler, D. Rojatz, V. Simovska, and R. Forster, "Effects of student participation in school health promotion: a systematic review," Health Promotion International, vol. 32, pp.195–206, 2017.
[12] B. R. Belland, C. M. Kim, and M. J. Hannafin, "A framework for designing scaffolds that improve motivation and cognition," Educational Psychologist, vol. 48, n. 4, pp.243–270, 2013.
[13] J. Van de Pol, M. Volman, and J. Beishuizen, "Scaffolding in teacher–student interaction: a decade of research," Educational Psychology Review, vol. 22, n. 3, pp. 271–296, 2010.
[14] R. Goldstone and J. Son, "The transfer of scientific principles using concrete and idealized simulations," Journal of the Learning Sciences, vol. 14, n. 1, pp. 69-110, 2005.
[15] M. J. Silva, J.B. Lopes, and A. A. Silva, "Using senses and sensors in the environment to develop abstract thinking – a theoretical and instrumental framework," Problems of Education in the 21st century, vol. 53, pp. 99-119, 2013.
[16] Y. Rogers, K. Connelly, W. Hazlewood, and L. Tedesco, "Enhancing learning: a study of how mobile devices can facilitate sense making," Personal and Ubiquitous Computing, vol. 14, n.2, pp.111-124, 2010.
[17] M. J. Silva, et al., "Adding space and senses to mobile world exploration," in Mobile Technology for Children, A. Druin (ed.), pp. 147-170. Boston: Morgan Kaufmann, 2009.
[18] M. J. Silva, J. C. Lopes, P. M. Silva, and M. J. Marcelino, "Sensing the schoolyard: using senses and sensors to assess georeferenced environmental dimensions," ACM 1st International Conference and Exhibition on Computing for Geospatial Research & Application, ACM, 2010.
[19] D. S. Fraser, et al.. "The SENSE project: a context-inclusive approach to studying environmental science within and across schools," Proceeding of CSCL'05, edited by International Society of the Learning Sciences, pp.155–9. Taiwan: Taipei, 2005.
[20] A. Zucker, R. Tinker, C. Staudt, A. Mansfield and S. Metcalf, "Increasing science learning in grades 3-8 using computers and probes: findings from the TEEMSS II project," Proceedings of the NARST 2007 Annual Meeting, 2007.
[21] E. van den Berg, F. Schweickert, and R. van den Berg, "Science, sensors and graphs in primary schools", Proceedings of the GIREP Conference 2010, pp. 1-9, 2010.
[22] S. Ananthanarayan, K. Siek, and M. Eisenberg "A craft approach to health awareness in children," Proceedings of the 2016 ACM Conference on Designing Interactive Systems, pp. 724-735, 2016.
[23] J. B. Lopes, J. P. Cravino, M. Branco, E. Saraiva, and A. A. Silva, "Mediation of student learning: dimensions and evidences in science teaching," PEC 2008—Problems of Education in the 21st Century, vol. 9, n. 9, pp.42–52, 2008.

Effect of Communicable Disease on the Performance of Elementary School Students

A Case Study of Leishmaniasis among Syrian Refugees in Some Bekaa (Lebanon) Area Schools

Firyal J. Zayyoun
School of Education
Lebanese International University (LIU)
Nabatieh, Lebanon
Email: 30831028@students.liu.edu.lb

Ehssan Sharif-Askari, Wiam Ramadan, Nancy Hallal
School of Arts and Sciences
International University of Beirut (BIU)
Beirut, Lebanon
Email: {ehssan.askari; wiam.ramadan; nancy.hallal}@b-iu.edu.lb

Hassan M. Khachfe
Business, Educational, & Medical Optimization Research Institute (BE-MORE)
Lebanese International University (LIU)
Nabatieh, Lebanon
Email: hassan.khachfe@liu.edu.lb

***Abstract*– It is well known that education in a child life has the power to develop him/her personally, socially, as well as economically. Due to the value of education and its significance, governments, Non-Governmental Organizations (NGOs), as well as educational institutions are investing more than ever before on education. In order to bring improvements to this sector, some of the factors influencing the academic performance of the students such as attitude of the students, the leadership aspects, and psychological and health related factors are being studied. Since the Syrian conflict started, Lebanon became a common destination receiving a huge number of Syrian refugees that are living in camps spread all over the country, with the largest concentration in the Bekaa Valley. Generous steps are being taken to increase the access to formal education, such as offering free public education and opening second shifts in the afternoon. Yet barriers, such as child labor and health related factors, like the spreading of the communicable disease leishmaniasis are keeping children out of classroom. The present study was done with the aim of investigating the effect of leishmaniasis on the performance and the academic achievement of Syrian children. The results showed varying degrees of knowledge and dealing with the case of leishmaniasis. The disease clearly had an effect on the students' attendance in schools, and by proxy on their academic performance.**

Keywords–communicable diseases; leishmaniasi; Leishmania; attendance; absenteeism; Syrian refugees.

I. INTRODUCTION

Since the eruption of the Syrian conflict, Lebanon began hosting fleeing refugees. To date, the country has sheltered around 1.5 million Syrian refugees, as estimated by the Lebanese government and a number of local NGOs. According to a report published by the United Nations High Commissioner for Refugees (UNHCR) in 2016, the refugees were spread all over the country as follows: Bekaa (36%), Beirut (26%), North Lebanon (26%), and South Lebanon (12%) [1]. Unfortunately, Syrian refugees live under very poor conditions. Inadequate sanitation, lack of access to clean water, overcrowding in their temporary settlements as well as limited access to healthcare infrastructure favor the spreading of communicable diseases, such as leishmaniasis. Leishmaniasis has emerged in Lebanon, but it has been mainly contained to refugee populations. Young people usually are the most affected due to their lack of previous exposure to the disease. The present study aimed at assessing the effect of leishmaniasis on the academic performance of elementary school students. The data was collected through a survey identifying the type of infection, the level of academic performance, the awareness among school's administration, teachers and parents. In addition, the negative impact of infection on the students' performance due to their absence from the class was investigated by directly interviewing the patients. The data clearly revealed that leishmaniasis has a direct effect on students' academic performance. This effect is in particular due to their absence from schools due to infection.

The remainder of this paper is organized as follows: Section II introduces the topic. Section III describes the methodology used. Section IV presents the results. Section V discusses the results obtained. The conclusion and acknowledgement close the article.

II. LITERATURE REVIEW

Leishmaniasis is an endemic disease in Syria, and the presence of war has greatly increased the risk for new cases in Syria as well as in the surrounding areas of the Middle East. The population displacement, poverty, and the poor living conditions, among other factors, help in the rapid transmission of the infection [2] [3]. Following the civil war in Syria, the Syrians started migrating to Lebanon and living in Syrian camps on the border with Syria, where the highest prevalence of leishmaniasis is recorded. School students are one of the main populations that are affected by this problem [4]-[6].

A. *Communicable diseases*

Communicable diseases are a main cause of morbidity and mortality caused by microorganisms in human beings [7] [7]. The risk of communicable disease is related to the characteristics of the affected population, such as malnutrition, host immune response, the access to health care services, and availability of safe clean water. Leishmaniasis is a parasitic zoonosis caused by protozoans of the genus Leishmania transmitted by insects known as phlebotomines sand flies, which are found in wild or urban environments [8]. Transmission of the disease occurs mainly through the infected insects; non-vector transmission is rare. The Leishmania protozoan was first described by Leishmania and Donovan in 1903 [9] and more than 50 Leishmania species have been identified worldwide; at least 21 of these species have significant medical importance. This parasite is endemic in at least 98 countries, including the eastern Mediterranean region [10].

There are three main forms of leishmaniasis: visceral leishmaniasis (VL), cutaneous leishmaniasis (CL) and mucocutaneous leishmaniasis (ML). CL is the most common form of leishmaniasis, characterized by skin manifestation causing skin lesions, papules, and nodules that may ulcerate leaving scars and serious disability. Despite a wide distribution, around one third of CL cases occur in the Americas, the Mediterranean basin, mainly in Iraq, Afghanistan, Iran, and Syria [10]. Mucocutanueous leishmaniasis is the result of untreated lesions of cutaneous infection causing total or partial destruction of mucus membrane of mouth, throat, and nose, and may spread to the adjacent mucosal surface. The most severe form of leishmaniasis is VL that is potentially fatal in over 95% of cases, if left untreated. VL affects internal organs including kidney, bone marrow, liver and spleen, causing enlargement in these organs, weight loss, fever and anemia. The epidemiology of these types of leishmaniasis depends on the characteristics of each parasite species [11].

Even though the risk of leishmaniasis is increasing worldwide due to many factors, they are gaining lesser attention than other communicable disease such as Malaria and AIDS [12]. Every year, 1.3 million new cases are reported with 20000 to 40000 cases of death [12]. According to the World Health Organization (WHO), the major risk factors of leishmaniasis are socioeconomic conditions, where poverty increases the risk of infection due to lack of waste management, little access to health care services, and open sewerage. These conditions may increase the breeding of female sand fly. In addition, lacking of different proteins, some vitamins and minerals due to undernourishment have been shown to increase the risk of VL [11]. Moreover, climate change such as change in temperature, rainfall and humidity affect strongly the distribution of vectors and reservoir hosts and their survival. More importantly, population mobility and displacement of non-immune people into areas with existing transmission cycles will increase the spread of the infection [11].

According to Alawieh *et al.* [3], no vaccine for Leishmania is available even though vaccination remains the most promising approach for the prevention. Nevertheless, the main way to prevent leishmania infection is by limiting exposure to the vector through insect repellent, insecticide, and controlling sand flies multiplication [4]. Another efficient method of prevention is through the establishment of a control program, awareness campaign for the community, in addition to increase coordination among the agencies [11]. Early diagnosis helps in reducing the prevalence of leishmeniasis and its spreading. The diagnosis of different forms of leishmaniasis is based on clinical manifestation with serological tests and it is type dependent [4].

B. Communicable disease among Syrians

Leishmania is one of the most important protozoan infections in Middle East and North Africa region [4]. One of the endemic countries of CL is Syria [13]. 90% of CL in Syria are caused by *Leishmania tropica* and the remaining cases are caused by *Leshmania infantum* [13]. Since Syria has the highest prevalence of CL mainly in Aleppo, leishmaniasis is known there as “Aleppo boil” [4]. The socioeconomic and environmental factors in 1990 caused the rapid spreading of CL in Syria [14]. During this time, the Syrian people migrated from rural to urban areas, thus, resulting that people with no or low immunity became highly exposed to CL. Despite the governmental effort in controlling sand fly multiplication through insecticidal treatment in 1991 [14], the CL incidence in Syria began to rapidly rise again. The increase in CL incidence was mainly due to the inability of municipal departments to provide adequate health care services, sanitation hygiene, and insecticide treatment [15].

C. Student performance in relation to class attendance

Education is a special process that improves skills and knowledge of individual to be able to participate in social activities and to promote and empower a community. However, Sintayehu Mekonnen in his study conducted to examine problems that confronted academic performance [5] concluded that the end result of educational process is challenged by different factors that may affect high quality of academic performance of learners.

Different studies explored factors affecting student's performance and student's achievements [16]-[18]. In a study published in 2011 by Saleem and Qureshi to identify different factors affecting student's achievements in higher education revealed that achievements are directly affected by academic factors, personal factors and socio-economic factors [16].

In addition, the academic performance of individuals is strongly affected by the type of education and the ability to provide the appropriate environment to accommodate individuals learning and educational needs. Karemera *et al.* [19] found that technology including computer lab, and the quality of library in institutions are directly related to students' performance. The study also revealed that student's achievement is positively affected by the service received and concluded that academic performance and school achievement are not associated with family income level. Furthermore, a study by David Romer confirmed that absenteeism and entry qualification affect learning; attendance should be mandatory [20]. In addition, the study found that environmental and motivational factors have a moderate effect in student academic performance.

In accordance with the above study emphasizing the effect of absenteeism on student's performance, Harb and El-

Shaarawi examined the effect of family size, gender, and place of living on students' performance [18]. They concluded that the most important factor affecting students' academic performance is class participation and students competence in English [18].

Shehry and Youssif in their assessment of factors affecting student's performance [21] attributed the bad academic performance to the long distance between schools and home. In addition to repeated absence, arriving home late and tired, the inability to complete and concentrate on homework, were additional factors that affected significantly students' performance [21].

In another study published by Hijazi and Naqvi, the authors have investigated different factors affecting student's performance, such as mothers' age, mothers' education, family income, study hours, and class attendance percentage. The data collected by authors showed clearly that class attendance is one of the main factors that have a positive effect on student's performance; regular attendance in college does contribute to a student's performance [22]. In addition, the authors found that educated mothers significantly affected their children performance, as compared to non-educated mothers, by keeping proper check on kids' activities [22].

Osaikhiuwu investigated, in his study, the institutional factors affecting students' performance, such as overcrowded lecture rooms, students' relationships, water supply, physical environment, and conditions of the facilities in higher educational institutions [17]. The author found that facilities in the system and favorable learning environment including water and electricity supply enhance students' performance. These findings are consistent with Karemeras' study [19] where he found that students' academic performance significantly correlated with overcrowded lecture rooms that cause unfavorable learning environment. In addition to institutional factors, other factors such as self-motivation, socioeconomic status, students' age, and some others could play a major role in determining students' performance [5].

III. Methodology

For this study, we use and apply a descriptive mixed method. Creswell defines the term mixed method as a research approach that arises from merging both qualitative and quantitative methods in a constant long term program of investigation to address their research questions. Such integration allows more harmonious usage of data than selectively using each type alone [23]. According to the author, using mixed study has several advantages:

1. In a mixed method, we confirm our quantitative results using qualitative experiment. As a result, this method will enable the researcher to relate and confirm the effect of leishmaniasis on the performance of elementary school students, by using the data from both an interview (qualitative), as well as a survey (quantitative), in the Bekaa schools.
2. The mixed method gives us multiple perspectives and more understanding of the topic by having results not only from Bekaa schools but also from directly interviewing parents of infected students at refugee camps.

This study is classified as descriptive mixed method as it analyses data collected from survey, interviewing teachers/ parents of infected students and mangers at Bekaa schools.

A. *Data collection*

Primary approval from the University's Institutional Review Board (IRB) was granted to conduct research at Bekaa schools, where students are potentially infected by leishmaniasis. The surveys were distributed, and the data were collected by hand, to ensure that the queried people understood the questions: an important aspect due to the social circumstances in which the study was conducted. Also, personal interviews with Bekaa school principals, the head of Communicable Diseases Department at the Lebanese Ministry of Public Health (MoPH), some NGOs, and managers of drug dispensaries in Western Bekaa were also conducted.

B. *Data analysis*

For quantitative data, all statistical analysis was done using the statistical package SPSS [24]. Analysis of the Variance (ANOVA) was used to see if there is a relation between student absence and academic achievements. For qualitative data, the interviews were analyzed.

IV. Results

Ninety percent of the surveyed schools reported having Syrian refugees among their student bodies. 83.6% of these schools testified that they have formal medical supervision of the students. What was alarming, however, was that only 20% of the schools had a formal awareness campaign about communicable diseases. While 4% of the schools had no form of health awareness, 53% did have some scattered activities in that regard, ranging from seminars, to weekly sessions, to some publications, with the different schools approaching the matter differently. Such a fragile awareness affected – of course – the student knowledge about the disease. Only 27% answered correct queries about leishmaniasis, whereas 20% had completely wrong answers, and the rest showed some degree of knowledge. The above information becomes somehow important when the sample showed that 97% of the interviewed reported encountering at least one case of leishmaniasis.

In terms of gender, the percentages of affected boys and girls were 60.3 and 39.7, respectively. The reported cases ranged from 3 to 14 years of age, with 52% of the cases in the 6-10 years bracket, 7% in the upper age (10-14 yrs) and 7% in the lower age (3-5 yrs) brackets. 93% of the reported cases were of the type "visceral leishmaniasis".

The academic performance of the students under survey was probed prior, during, and after being diagnosed with – and later on treated from – leishmaniasis. The data is presented in Table I. Next, the effect of attendance was investigated, and the results are shown in Table II.

As for the parents of affected students, 75.9% of them received general health awareness, 3.4% received published material specific to the disease, and 3.4 % were trained to deal with such cases, whereas some 17.2 % showed complete indifference. It was worth noting that 80% of the institutions (schools, NGOs, etc.) dealing with the above cases provided access to social workers that were trained to deal with both the infected students, their parents, and their teachers.

TABLE I. THE ACADEMIC PERFORMANCE OF THE STUDENTS IN THE DIFFERENT PHASES OF THE INFECTION.

	Weak	Acceptable	Good	Very Good	Excellent
Pre-infection	0 %	10.3 %	75.9 %	6.9 %	6.9 %
During infection	55.2 %	34.4 %	10.4 %	0 %	0 %
Post-infection	24.2 %	41.4 %	27.6 %	3.4 %	3.4 %

V. DISCUSSION

The Lebanese health system is composed of services rendered by the public and the private sectors, as well as the nongovernmental organizations' sector. The dominant sector is the private one, since it provides most of the services [25]. In addition, there is a large number of primary health care centers that provide health services, including vaccination, management of communicable diseases, and medicines [25]. Around 50% of Syrian refugees live in the poorest areas of Lebanon informally [25]. According to an assessment conducted by UNHCR in 2016, 310,000 individuals among the Syrian refugees in Lebanon received primary healthcare, and 98,861 individuals benefitted from several lifesaving and obstetric healthcare services. According to Reidner and Sabbah, "limited funds are available for ensuring equitable provision of health services to meet essential health needs at the primary, secondary and tertiary health care levels." Consequently, in the six years of the Syrian crisis, the healthcare services rendered across the Syrian refugee population in Lebanon remain a serious concern [25].

TABLE II. THE ATTENDANCE OF THE STUDENTS IN THE DIFFERENT PHASES OF THE INFECTION.

	Extremely low	Lower than Normal	Normal*	Permanent‡
Pre-infection	0 %	0 %	62.1 %	37.9 %
During infection	6.9 %	72.4 %	20.7 %	0 %
Post-infection	6.9 %	41.4 %	44.8 %	6.9 %

* Normal attendance was based on the average attendance of no-infected students in the same class

‡ No absence was reported during the timeframe of the study

The wide distribution of Syrian refugees in Lebanon caused sudden increase in leishmania cases, where this new outbreak of leishmaniasis in Lebanon is the first of its kind for more than a decade [4]. According to [4], the Lebanese Ministry of Public Health (LMoPH) implemented an active surveillance for leishmaniasis since the beginning of the Syrian conflict; LMoPH diagnosed the cases by traditional techniques of smear, histological analysis of skin biopsies and parasite culture [4]. The USA and several developed countries put together collaborative efforts to prevent the globalization of this disease through its transmission by troops, travelers, immigrants, NGO workers and tourists visiting or passing through countries with endemic leishmaniasis in the Middle East [4]. According to Tokajian et al. [2], leishmaniasis is considered one of the most neglected diseases in the world, in the time where the parasite may be able to propagate in different areas in Lebanon, particularly in Bekaa district, where the vector may find a favorable habitat and where appropriate zoonotic reservoirs are present [4].

The current study was conducted solely on students aged 3-14 years, and who were – at the time of the data collection – matriculated in some West Bekaa area schools. The results showed varying degrees of knowledge and dealing with the case of leishmaniasis. The disease clearly had an effect on the students' attendance in schools, and by proxy on their academic performance.

Part of the causes of such variations in dealing with the problem is due to the political strings attached to the different Lebanese ministries (e.g., Public health, Social affairs, Refugees and Displaced, Education, etc.), and the conflicts that may exist between the heads of these ministries. A clear indication to such an argument was the lack of coordination between the schools (under the control of the Ministry of Education), the NGOs (under the control of the Ministry of Social Affairs), and the Primary Healthcare Centers (under the control of the Ministry of Public Health). Our study reported that 64.3 % of the schools lacked any kind of coordination with the MoPH centers, whereas on 28.6 % of the schools were in direct relationship with the MoPH offices. Such an argument is further corroborated by the fact that the MoPH provided the necessary medicines and vaccines to 70% of the reported cases.

VI. CONCLUSION

More evidence is accumulated that refugee camps are not good for anyone, especially for kids. In this study, we addressed one communicable disease; however, there are still more issues that will remain to be answered. These factors include – but are not limited to – psychological stresses during and post-migration, challenges that affect the ability of the refugee students to adapt to the host schools system, etc. This is despite the fact that educational access and opportunities for camps refugee children remains low. In conclusion, our study raises a critical point of the effect of communicable diseases on the performance of elementary students, in particular the Syrian refugees. Leishmaniasis is an example of one factor that keeps children out of the classroom. How other health related factor may also affect could be further evaluated. In addition, public awareness within Syrian refugees might also be a subject of further investigation to assess its effect on children education. This can be further implemented by school administration and their partnership with parents and/or guardian of the refugee students. The main question that remains to be answered with respect to the Lebanese community is whether the Lebanese authorities have enough knowledge in dealing with such education-related health problems despite the fact that leishmaniasis is not endemic in Lebanon.

ACKNOWLEDGMENT

The authors would like to thank all participating schools in the West Bekaa District, the Head of the Communicable Diseases Department at the MoPH, Dr. Atika Berry, and the Officers at the Beyond Association in Nabatieh.

REFERENCES

[1] UNHCR , "UNHCR Lebanon: back to school," 8 January 2016. [retrieved: October, 2018]. Available: www.data.unhcr.org.

[2] T. Salloum, I. Khalifeh, and S. Tokajian, "Detection, molecular typing and phylogenetic analysis of Leishmania isolated from cases of Leishmaniasid among Syrian refugees in Lebanon," Parasite Epidemiology and Control, pp. 159-168, 2016.

[3] M. Douba, A. Mowakeh, and A. Wali, "Current status of cutaneous leishmaniasis in Aleppo, Syrian Arab Republic," Bulletin of the World Health Organization, pp. 253-259, 1997.

[4] A. Alawieh et al., "Revisiting leishmaniasis in the time of war: the Syrian conflict and the lebanese outbreak," International Journal of Infectious Diseases, pp. 115-119, 2014.

[5] S. Mekonnen, "Problems Challenging the Academic Performance of Physics Students in Higher Governmental Institutions in the Case of Arbaminch, Wolayita Sodo, Hawassa and Dilla Universities," Scientific Research, pp. 362-375, 5 March 2014.

[6] N. Khalid and K. Mehmood, "Effect of Absenteeism on Students Performance," International Journal of Scientific and Research Publications, pp. 151-168, September 2017.

[7] World Health Organization, "Disaster Risk Management for Health, "Communicable Diseases," Global Platform, 2011.

[8] E. Petersen, S. Baekeland, Z. A. Memish, and H. Leblebicioglu, "Infectious disease risk from the Syrian conflict," International Journal of Infectious Diseases, pp. 666-667, June 2013.

[9] K. Bathena, "A Mathematical model of cutaneous leishmaniasis," 1 June 2009. [retrieved: October, 2018]. Available: www.scholarworks.rit.edu.

[10] M. Saroufim et al., "Ongoing Epidemic of Cutaneous Leishmaniasis among Syrian Refugees, Lebanon," Emerging Infectious Diseases, pp. 1712-1714, 2014.

[11] World Health Organization, "World Health Organization:Media Centre," April 2017. [retrieved: October, 2018]. Available: www.who.int.

[12] N. Salam, W. M. Al-Shaqha, and A. Azzi, "leishmaniasis in the middle east: incidence and epidemiology," PLOS Neglected Tropical Diseases, pp. 1-6, 2014.

[13] N. Haddad, H. Saliba, A. Altawil, J. Villinsky, and S. Al-Nahhas, "Cutaneous leishmaniasis in the central provinces of Hama and Edlib in Syria: Vector identification and parasite typing," Parasites & Vectors, 2015.

[14] R. Du, P. Hotez, W. Al-Salem, and A. Acosta-Serrano, "Old World Cutaneous Leishmaniasis and Refugee Crises in the Middle East and North Africa," PLOS Negleceted Tropical Diseases, pp. 1-11, 26 May 2016.

[15] R. Du, P. J. Hotez, W. S. Al-Salem, and A. Acosta-Serrano, "Old World of Cutaneous Leishmaniasis and Refugee Crises in the Middle East and North Africa," PLOS Neglected Topical Diseases, pp. 1-11, 26 May 2016.

[16] M. A. Saleem and M. I. Qureshi, "Credentials and examination of the factors affecting the students' academic achievement in higher education (A case study of universities in public and private sector at D.I.Khan), Dera Ismail Khan, Gomal University Journal of Research, 27(2), pp. 74-80, December, 2011.

[17] O. C. Osaikhiuwu, "Institutional Factors Affecting the Academic Performance of Public Administration Students in a Nigerian University," Canadian Center of Science and Education, pp. 171-177, 29 October 2014.

[18] N. Harb and A. El-Shaarawi, "Factors Affecting Students' Performance," Munich Personal RePEc Archive, p. paper No. 13621, 2006.

[19] D. Karemera, L. J. Reuben, and M. R. Sillah, "The Effects of Academic Environment and Background Characteristics on Student Satisfaction and Performance: The Case of South Carolina State University's School of Business," June 2003. [retrieved: October, 2018]. Available: www.questia.com/read/1G1-103563753/the-effects-of-academic-environment-and-background. [Accessed June 2003].

[20] D. Romer, "Do Students Go to Class? Should they?," The Journal of Economic Perspectives, pp. 167-174, 1993.

[21] A. A. Shehry and S. M. A. Youssif, "Factors affecting academic performance of undergraduate students at Najran preparatory year for girls-Najran Unuversity 2015-2016," International Journal of Asia Social Science, pp. 1-18, 2017.

[22] S. T. Hijazi and S. R. Naqvi, "Factors affecting students' performance," Bangladesh e-Journal of Sociology, vol. 3, pp. 1-10, January 2006.

[23] J. W. Creswell, "Steps in Conducting a Scholarly Mixed Methods, DBER Speaker Series. 48, 2013.

[24] M. Norusis, "SPSS 17.0 guide to data analayis," Upper Saddle River, NJ: Prentic Hall, 2009.

[25] G. Reidner, R. Sabbah, and C. Diaz-Herrera, "WHO support to the humanitarian response in Lebanon," 2014-2015. [retrieved: October, 2018]. Available: www.limelightprod.com.